SECOND EDITION

Storey's Guide to
RAISING
DUCKS

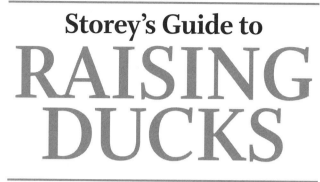

Breeds . Care . Health

DAVE HOLDERREAD

Storey Publishing

*The mission of Storey Publishing is to serve our customers by
publishing practical information that encourages
personal independence in harmony with the environment.*

Edited by Sarah Guare and Deborah Burns
Art direction and book design by Cynthia N. McFarland
Cover design by Kent Lew
Text production by Jennifer Jepson Smith

Cover photograph by © Jowita Stachowiak/iStockphoto
Author's photograph by © Christopher David Miller Green
Interior photography by Dave Holderread
Illustrations by © Elayne Sears

Indexed by Christine R. Lindemer, Boston Road Communications

Storey Publishing
210 MASS MoCA Way
North Adams, MA 01247
www.storey.com

Printed in the United States by Versa Press
10 9 8 7 6 5 4 3 2 1

LIBRARY OF CONGRESS CATALOGING-IN-PUBLICATION DATA

Holderread, Dave.
  Storey's guide to raising ducks / by Dave Holderread. — 2nd ed.
    p. cm.
  Includes index.
  ISBN 978-1-60342-692-3 (pbk. : alk. paper)
  ISBN 978-1-60342-693-0 (hardcover : alk. paper)
  1. Ducks. I. Title.
SF505.H65 2011
636.5'97—dc22
                                                        2010030074

STOREY'S GUIDE TO RAISING DUCKS

Contents

Acknowledgments

This book is possible because a host of people have shared their time, knowledge, and expertise over the years. Because a significant portion of the information contained herein was gained from research and observations made here at Holderread Waterfowl Farm & Preservation Center (HWF&PC), I am indebted to everyone who has helped keep this labor of love functioning. There is not room to mention everyone, but I want to acknowledge the following people for their key roles.

My parents, Rachel and Wilbur Holderread, who encouraged me throughout my childhood to pursue my fascination with all things feathered, and gave me a foundation in the art and science of animal husbandry.

Frank Braman, a talented poultry breeder and judge who took me under his wing in my early teens and taught me selection and breeding methods we use to maintain and improve the quality of the waterfowl raised here at HWF&PC.

D. Phillip Sponenberg, DVM, PhD, who has encouraged us in our work with endangered breeds and has been a valued resource for helping puzzle through genetics questions.

Carol Deppe, geneticist, plant breeder, and advocate for the integration of ducks and gardens, who challenged me to develop a uniform nomenclature for duck color genetics.

David Omick and Pearl Mast, who lend their design and building expertise, and are part of the advisory team of HWF&PC.

The Miller Green Family: Dale, who pinch-hits at chore time when needed and helps with technical support. Wanita, who has helped in all facets of HWF&PC for over 30 years — from chores to webmaster to finding the photograph for the front cover of this book. Christopher and Maggie, who have been an integral part of our waterfowl program since they could walk. Christopher's keen eye for simplifying tasks and Maggie's devotion to helping injured animals heal are inspiring.

And most of all to Millie, my wife, who has been my biggest supporter and cohort as I have pursued my waterfowl passion. From daily farm chores to keeping volumes of records to editing my writing — this book would not have been possible without her numerous talents and willingness to pitch in wherever needed.

Foreword

The Importance of Ducks — and This Book!

Ducks are the most underappreciated of all domesticated poultry and livestock species. It is hoped that this will change as more people learn about ducks, their characteristics, and their incredible productivity. This book provides wonderful coverage of both the basics and the details of duck production. Dave and Millie Holderread have spent many decades in working with ducks, and no other people are more qualified to pass along accurate and useful information about ducks.

Ducks, with their incredible array of different breeds, are important and useful resources for agriculture, yet also offer breeders much enjoyment. Each duck breed comes to us from a unique combination of place, people, and breeding goals, and each is well presented in this book.

Dave understands the intricacies of color genetics incredibly well and has shared that wealth here. He also sheds light on the fine points of breeding ducks for high levels of production — both for eggs and for meat. The approaches in this book will ensure that ducks remain productive, hardy, beautiful, and enjoyable. Dave also clearly explains the finer points of selecting breeding birds for the production of great show birds.

The threat of losing distinct and productive breeds of ducks looms ever larger as the years go by. Conservation is ideally accomplished by lots of breeders raising lots of ducks, and selecting them for excellence in production, adaptability, soundness, and form. The approaches outlined here will enable the next generation of duck breeders to make great strides in the conservation of the genetic treasures that are contained in each breed package.

This book is a deep and useful resource for anyone from beginner to advanced breeder and is jam-packed with information available nowhere else. This includes recent scientific understandings and many insights and "tricks" from old-time breeders that are sure to contribute to success.

D. PHILLIP SPONENBERG, DVM, PHD
Professor of Pathology and Genetics, Virginia Tech
Technical Advisor, American Livestock Breeds Conservancy

1

Why Ducks?

THE POPULARITY OF DUCKS — often described as the happiest animals in the barnyard — is increasing in many areas of the world. It appears that the rest of us are beginning to understand what many Asians and Europeans have known for centuries — ducks are one of the most versatile and useful of all domestic fowl. For many circumstances it is difficult to find a better all-purpose bird than the duck.

Duck Attributes

There are many reasons why people raise ducks. These amazingly adaptable fowl produce meat and eggs efficiently, in many situations require a minimum of shelter from inclement weather, are active foragers, consume large quantities of flies, mosquito larvae, and a wide variety of garden pests (such as slugs, snails, and grubs) and weed seeds, produce useful feathers, and are exceptionally healthy and hardy. A wonderful bonus to their myriad practical qualities is the entertaining antics and beauty they add to our lives.

Easy to Raise

People who have kept all types of poultry generally agree that ducks are the easiest domestic birds to raise. Along with guinea fowl and geese, ducks are incredibly resistant to disease. In many situations chickens must be vaccinated for communicable diseases and regularly treated for worms, coccidiosis, mites, and lice to remain healthy and productive. Duck keepers can often forget about these inconveniences. Even when kept under less than ideal conditions, small duck flocks are seldom bothered by sickness or parasites.

Resistant to Cold, Wet, and Hot Weather

Mature waterfowl are practically immune to wet or cold weather and are profoundly better adapted to cope with these conditions than are chickens, turkeys, guineas, or quail. Thanks to their thick coats of well-oiled feathers, healthy ducks of most breeds can remain outside in the wettest weather. Muscovies (which tend to have less water repellency than other breeds) and any duck that has poor water repellency due to infirmity should have free access to dry shelter during cold, wet weather.

While chickens have protruding combs and wattles that must be protected from frostbite, as well as bare faces that allow the escape of valuable body heat, ducks are much more heavily feathered and are able to remain comfortable — if they are provided dry bedding and protection from wind — even when the temperatures fall below 0°F (–18°C). (When people express concern about mature, healthy ducks being cold, I remind them that these waterfowl have the original and best down coats on the market!) Because ducks have the ability to regulate how much down they grow depending on weather conditions, they also thrive in hot climates if they have access to plenty of shade and cool drinking water. During torrid weather bathing water or misters can be beneficial.

Insect, Arachnid, Snail, and Slug Exterminators

Because they nurture a special fondness for mosquito pupae, Japanese beetle larvae, potato beetles, grasshoppers, snails, slugs, flies and their larvae, fire ants, and spiders, ducks are extremely effective in controlling these and other pests. In areas plagued by grasshoppers, ducks are used to reduce plant and crop damage during severe infestations. Where liver flukes flourish, ducks can greatly reduce the problem by consuming the snails that host this troublesome livestock parasite.

Under many conditions two to six ducks per acre (0.5 ha) of land are needed to control Japanese beetles, grasshoppers, snails, slugs, and fire ants. To eliminate mosquito pupae and larvae from bodies of water, provide six to ten ducks for each acre of water surface. With observation and experience you will be able to determine the number of ducks needed for your specific situation. The breeds of ducks in the Lightweight class and the larger bantam breeds are the most active foragers, making them the best exterminators for large areas. However, as noted in the chart of breed profiles on pages 28–29, most other breeds are also good foragers.

GENERAL COMPARISON OF POULTRY

Bird	Raisability	Disease Resistance	Special Adaptations
Coturnix quail	Good	Good	Egg and meat production in extremely limited space
Guinea fowl	Fair–good	Excellent	Gamy-flavored meat; insect control; alarm. Thrive in hot climates.
Pigeons	Good	Good	Message carriers; meat production in limited space. Quiet.
Chickens	Fair–good	Fair–good	Eggs; meat; natural mothers. Adapt to cages, houses, or range.
Turkeys	Poor–fair	Poor–fair	Heavy meat production
Geese	Excellent	Excellent	Meat; feathers; lawn mowers; "watchdogs"; aquatic plant control. Cold, wet climates.
Ducks	Excellent	Excellent	Eggs; meat; feathers. Insect, snail, slug, aquatic plant control. Cold, wet climates.

Productive

Ducks are one of the most efficient producers of animal protein. Strains that have been selected for high egg production (especially Campbells, Welsh Harlequins, and special hybrids) lay as well as or better than the best egg strains of chickens, averaging 275 to 325 eggs per hen per year. Furthermore, duck eggs are 20 to 35 percent larger than chicken eggs produced by birds of the same size. Unfortunately, in many localities, strains of ducks that have been selected for top egg production are not as readily available as egg-bred chickens.

Meat-type ducks that are raised in confinement and fed an appropriate diet are capable of converting 2.6 to 2.8 pounds of concentrated feed into 1 pound of bird. When allowed to forage where there is a good supply of natural foods, they have been known to do even better. The only domestic land animal commonly used for food that has better feed conversion is the industrial hybrid broiler chicken, with a 1.9:1 ratio.

Excellent Foragers

Ducks are energetic foragers. Depending on the climate and the abundance of natural foods, they are capable of rustling from 15 to 100 percent of their own diet. Along with guinea fowl and geese, ducks are the most efficient type of domestic poultry for the conversion of food resources that normally are wasted — such as insects, slugs, snails, windfall fruits, garden leftovers, and weed plants and seeds — into edible human fare.

Aquatic Plant Control

Ducks are useful in controlling some types of unwanted plants in ponds, lakes, and streams, improving conditions for many types of fish. In most situations 15 to 30 birds per acre (0.5 ha) of water are required to clean out heavy growths of green algae, duckweed (*Lemna*), pondweed (*Potamogeton*), wid-

---

DUCKS IN THE JUNGLE

This story comes from John M. Jessup, RN, MPH, faculty member at Oregon Health and Science University. It demonstrates the truly global importance of waterfowl.

I was in Guatemala, working with the indigenous Kekchi people on ways to decrease the 50 percent–plus malnutrition rate among village children, when I realized how valuable ducks could be to the people there. Dave had previously told me about duck breeds that could forage for much of their food and still produce a lot of eggs pretty much year-round. The humid jungle environment was hard on chickens so we thought a pilot project with ducks might be a way to improve the diet of the families with malnourished children.

Dave and Millie donated 500 day-old Khaki Campbell and Magpie ducklings to our health program in the Peten Jungle. The Guatemalan government "taxed" us 250 ducklings to let them into the country. We drove 10 hours back to our village, where raised wooden cages with screen bottoms were waiting for them. Ducklings drink a lot of water with their food and before long we had a wet mess under the cages. I placed 1" × 8" boards around the bottom of the cage legs to contain the wet "mixture" of duckling droppings, feed, and water.

We planned to distribute the ducklings to Kekchi families when the ducklings were six weeks of age. After the first week, I noticed the "mixture" had a healthy population of maggots. Not one to let a good protein source go to waste, I made a screen dipper that allowed me geon grass (*Ruppia*), muskgrass (*Chara*), arrowhead (*Sagittaria*), wild celery (*Vallisneria*), and other plants that ducks consume. Once the plants are under control, 8 to 15 ducks per acre will usually keep the vegetation from taking over again.

In bodies of water containing plants submerged more than 2 feet (0.6 m), or when it is desirable to clean grass from banks, four to eight geese per acre of water surface should be used along with the ducks. (For geese to be effective, they must be confined to the pond and its banks with fencing.) In experiments conducted in Puerto Rico, waterfowl were not found to be effective in eliminating well-established infestations of tropical plants such as water

to extract and then wash the maggots, leaving a white wiggling mass of highly nutritious duckling feed. The maggots were then dumped into the ducks' feed troughs once or twice a day, until we distributed the birds to their excited new owners.

The ducks and eggs were very popular with the Kekchi. Both breeds thrived in the jungle environment. Over the next couple of years, we were able to improve the height and weight of the children we worked with. The Magpies were the most popular among the Kekchi. Surprisingly, the villagers reported that the Magpies produced more eggs than the Campbells on the village diet of cracked home-grown corn and whatever they could find foraging in the brush around their thatched huts.

The ducks were bred locally and sold by a nearly blind twelve-year-old Kekchi boy named Martin, who was unable to work in the corn and rice fields with his family. Martin developed a home business hatching and raising ducklings for sale. He initially used a hundred-egg incubator, and later, a flock of native Muscovies, to hatch the Campbell and Magpie eggs.

During my travels to other parts of Guatemala, I learned that Khaki Campbells and Magpies had been given to villagers through a government agriculture program. I suspect that these were some of the other 250 ducklings that the Holderreads had sent us.

spinach and water hyacinth. However, both of these plants will be eaten by ducks if chopped into small enough pieces, and can be used as a supplement to other feeds.

Garbage Disposal

Ducks are omnivores and will eat most food items that come from the kitchen or root cellar. The rule of thumb is this: If humans eat it, ducks most likely will also — as long as it is in a form they can swallow. They relish many kinds of leafy greens (they tend to be wary of red- or purple-colored leaves but, oddly, not fruit of these colors), garden vegetables and root crops, both temperate zone and tropical fruits (even bananas and citrus if they are peeled), canning refuse, most kinds of stale baked goods, and outdated dairy products and by-products such as cheese, whey, and curdled milk (these last two are best used to moisten dry foods, such as baked goods and finely ground feeds). To make it easier for these broad-billed fowl to eat firm vegetables and fruits, place apples, beets, turnips, and such on an old board and crush them with your foot or cut them into bite-size pieces.

Cooked potatoes can be an excellent source of carbohydrates and protein for ducks (avoid raw, green, or moldy potatoes; see page 234). Other root crops are typically consumed in larger quantities if cooked than if left raw.

Find creative ways of having your ducks utilize "waste" products, but avoid moldy or fermented foods and anything known to have harmful toxins, such as raw soybeans and potatoes. Keep in mind that certain sticky foods (such as milk) can compromise water repellency if allowed to splash onto the ducks' feathers.

During a 9-month research project in Puerto Rico, we supplied a flock of 40 mature Rouen ducks that were on pasture and had access to a 5-acre (2 ha) pond with nothing but leftovers from the school cafeteria (and oyster shells during the laying season). These "garbage-fed" birds remained in good flesh and showed no signs of poor health, although they produced 60 percent fewer eggs than a control group that was provided concentrated laying feed along with limited quantities of institutional victuals.

Useful Feathers

Down feathers come in a wide array of sizes, colors, and shapes and have both practical and artistic value. The down and contour body feathers of ducks are valuable as filler for pillows and as lining for comforters and winter clothing (see appendix E, Using Feathers and Down, page 335.) Fly fishers use duck feathers in fly tying, and artisans incorporate them into artworks. Because of their high protein content, the feathers yield valuable plant fertilizer when composted with other organic materials.

Valuable Manure

A valuable by-product of raising ducks is manure. Duck manure is an excellent organic fertilizer that is high in nitrogen. In some Asian countries duck flocks are herded through rice fields to eat insects, snails, and slugs and to pick up stray kernels of grain. The birds are then put on ponds where their manure provides food for fish.

Gentle Dispositions

Typically, ducks are not aggressive toward humans. Of the larger domestic birds, they are the least likely to inflict injury on children or adults. In the 50 years I have worked with ducks, the only injuries I've sustained have been small blood blisters on my arms and hands — received while attempting to remove eggs from under broody hens — and an occasional scratch when the foot of a held bird escaped my grasp. (There are exceptions. Factors such as the bird's personality, the environment in which it is raised, and the temperament

Ducks add interest and beauty to waterways and greatly reduce mosquitoes and other aquatic pests. Pictured are two Australian Spotted females on one of our ponds.

of its caretakers can alter the usual docility of the duck. For instance, many Muscovies interact well with humans, but due to their long claws and exceptional strength, the occasional aggressive one — usually male — can inflict painful injury.) If you do get scratched by the claws of a bird, prompt washing of the wound with hydrogen peroxide and the application of an antibiotic ointment will lessen the chance of infection.

Decorative and Entertaining

Along with having many practical attributes, ducks are beautiful and fun to watch as they enthusiastically go about their daily activities. Over the years many people have told me of the pleasure and relaxation they experience from the simple act of watching their ducks. Folks have given detailed accounts of setting up lawn chairs next to their duck yards or locating their pens within sight of the house, so their ducks can be observed from a picture window. A small flock of waterfowl can transform just another yard or pond into an entertainment center that provides hours of enjoyment.

I can still remember the first ducks I had as a young boy. Because our property had no natural body of water, I fashioned a small dirt pond in the center of the duck yard. After filling it with water, I watched as my two prized ducklings jumped in and indulged in their first swim. They played and splashed with such enthusiasm that it wasn't long before I was as wet as they were. And then — much to my delight — they began diving, with long seconds elapsing before they popped above the surface in an unexpected place. I was hooked, and I continue to be intrigued by the playfulness, beauty, and grace of swimming waterfowl.

Helping Conserve Rare Genetic Stocks

One of the cornerstones for long-term viability of food-producing systems is genetic diversity. Unfortunately, many of the old and most versatile breeds of ducks are in danger of being lost to extinction as "modern" agriculture practices rely almost exclusively on a few breeds that best tolerate being packed tightly together in climate-altering buildings. You can help conserve rare genetic stocks in a variety of ways. Breeding and dispersing rare breeds is an obvious choice. If you raise one or more of the endangered breeds, you are helping curb the dangerous erosion of genetic diversity. Even if you buy rare ducklings for butchering at maturity, you are helping the breed by creating demand for them.

COMPARISON OF DUCKS AND CHICKENS

Characteristic	Ducks	Chickens
Housing requirements	Minimal	Substantial
Height of fence to keep confined	2–3 feet (0.6–0.9 m)	4–6 feet (1.2–1.8 m)
Susceptibility to predators	High	Moderately high
Resistance to parasites and disease	Excellent	Fair
Resistance to hot weather	Good	Good
Resistance to cold weather	Good	Fair
Resistance to wet weather	Excellent	Poor
Foraging ability	Good–excellent	Fair–good
Scratching in dirt	None	Considerable
Probing in mud with bills or beaks	Considerable	None
Incubation period	28 days	21 days
Hatchability of eggs in incubators	65–90%	70–90%
Cost of day-olds in lots of 25	Ducklings double the price of chicks	
Starter feed — min. protein needed	16%	18%
Layer feed — min. protein needed	15%	15%
Age hens commence laying	16–24 weeks	18–24 weeks
Eggs laid per hen per year (wt.)	32–52 lbs (14.5–23.5 kg)	22–34 lbs (10–15.5 kg)
Feed to produce 1 pound of eggs	2.4–3.8 lbs (1–1.75 kg)	2.8–4.0 lbs (1.25–1.8 kg)
Part of diet hens can forage	10–25%	5–15%
Light required for top production	13–15 hours	14–17 hours
Annual mortality rate of hens	0–3%	5–25%
Efficient production life of hens	2–3 years	1–2 years
Protein content of eggs	13.3%	12.9%
Fat content of eggs	14.5%	11.5–12.5%
Cholesterol content of eggs	Ducks slightly higher	
Flavor of eggs	Similar; duck eggs some-times stronger	
Feed to produce 1 pound of bird	2.5–3 lbs (1–1.4 kg)	2–2.2 lbs (0.9–1 kg)
Age to butchering	7–12 weeks	7.5–16 weeks
Typical plucking time	3–15 minutes	2–10 minutes
Color of meat	All dark	Light and dark
Protein content of flesh	21.4%	19.3%
Fat content of carcass	16–30%	5–25%
Usefulness of feathers and down	Excellent	Fair

Some Points to Consider

Despite the unparalleled versatility and usefulness of ducks, there are several factors you should be aware of and understand. In some situations another type of fowl may prove more suitable.

Noise

Many people find the quacking of ducks an acceptable part of nature's choir. However, if you have close neighbors, the gabble of talkative hens may not be appreciated. Some breeds (and some individuals within a breed) are noisier than others. Typically, Call ducks are the noisiest, with Pekins being the second most talkative. Under many circumstances, a small flock consisting of any other breed will be reasonably quiet if not frightened or disturbed frequently. Muscovies are nearly mute, making them the least noisy of all breeds. Also, drakes of all breeds have weak voices, and for the control of slugs, snails, and insects in a town or suburb, they work fine.

Plucking (Picking)

Most people find that plucking a duck is more time-consuming than defeathering a chicken. But then, most of us who have had the privilege of dining on roast duck agree that it is time well spent. Furthermore, duck feathers are much more useful than chicken plumes. With good technique and experience, it is possible to reduce the picking time to 5 minutes or less.

Pond Density

Having large numbers of ducks on small ponds or creeks encourages unhealthy conditions and can result in considerable damage to bodies of water. One of the feeding habits of ducks is to probe the mud around the water's edge for grubs, worms, roots, and other buried treasures. A high density of ducks will muddy the water and hasten bank erosion. On the other hand, a reasonable number of birds (15 to 25 per acre [0.5 ha] of water) will improve conditions for fish, will control aquatic plant growth and mosquitoes, and will not significantly increase bank erosion.

Gardens

Ducks do an amazing job of controlling slugs, snails, and various types of harmful grubs and insects in gardens. They also will eat tender young grasses and broadleaf weeds that they find palatable. However, because they may also consume desired crops, to prevent the birds from doing more harm than good, follow these guidelines:

1. Don't let birds in until the crops are well started and past the succulent stage.
2. Keep ducks out when irrigating or when the soil is wet.
3. Fence off tender crops such as lettuce, spinach, cabbage, and green beans (most ducks can be kept out with a 24- to 30-inch [60 to 75 cm] barrier).
4. Remove birds when low-growing berries and fruits are ripe.
5. Limit the number of ducks to two to four adults for each 500 to 1,000 square feet (45 to 95 sq m) of garden space.

A method we have used successfully for decades is to pen ducks around the perimeter of the garden, where they intercept migrating slugs, snails, and insects. Then, when we are working in the garden during the growing season, we allow a few ducks into the garden to "vacuum up" the pests they are so adept at ferreting out. When we are ready to leave the garden for the day, the ducks are simply herded or enticed back to their garden-side enclosure with the feed can. During the nongardening season, the broad-billed exterminators are allowed into the garden daily.

Ducks can make a highly effective pest patrol in the garden, as long as you take care to keep them from tender plants and low-growing fruit.

Meat and Eggs

All types of poultry provide good food. However, there are variations in the flavor, texture, and composition of the meat and eggs produced by the diverse species. There are also differences in dietary needs and likes and dislikes of food among people. If you are seriously considering producing duck meat or eggs but have never cooked or eaten them, I recommend that you sample duck products before starting your flock. (This practice is wise before investing time and resources in any type of unfamiliar animal for food.) The following observations will help you evaluate your first encounter with duck cuisine.

1. The flavor of meat and eggs of any species is dramatically influenced by the animals' diet. Ducks whose diet includes fish or fish products or who are allowed to forage in areas containing strong-flavored plants will typically have strong or off-flavored meat and eggs, which are not typical of good duck products. (Remember: Ducks are true omnivores, which means they will eat a wider range of foodstuffs than most fowl. That said, in most situations even free-ranging ducks produce delightful eggs and meat.)

2. Ducklings that have been raised in close confinement and pushed for top growth rate are much fatter than ducks that have foraged for some of their own food and have grown at a slower pace, e.g., pastured ducks. The high fat content of quick-grown duckling makes its meat exceptionally succulent and provides valuable energy for persons who get strenuous physical exercise. However, most of us do not need that much fuel.

3. Duck eggs typically have a slightly higher cholesterol content than the average chicken egg. As with fat in meat, the amount of cholesterol in eggs is affected by the diet and lifestyle of the producing bird. Ducks that are active and forage for a portion of their diet typically produce eggs lower in cholesterol.

   When eggs are eaten in moderation, the difference in cholesterol between duck and chicken eggs probably is insignificant for people who get adequate exercise and eat sensibly.

4. Duck eggs are excellent for general eating and often preferred for baking purposes. When fried, the whites of duck eggs are often firmer than those of chicken eggs. While it is often said that duck eggs are unsatisfactory for meringues and angel food cakes, we have not found this to be true. In fact, one of our favorite treats for special occasions is whole-wheat angel food cake made with duck egg whites (see page 334).

2

External Features and Behavior

DUCKS ARE MASTERFULLY DESIGNED, down to the smallest detail, for both aquatic and terrestrial life. Being acquainted with the external features of ducks is a useful management tool and will deepen your appreciation for these incredibly versatile fowl.

Fundamental Duck Anatomy

Following are the basic anatomical features that you should know, and I suggest that you take a few minutes to familiarize yourself with the accompanying nomenclature diagram. It is important to remember that there are two distinct species of domestic ducks commonly raised: those breeds derived from the wild Mallard and Muscovies. These two species share some anatomical similarities but do have significant differences that I will point out when indicated.

Body Shape

For stability and minimal drag while swimming and diving, the lines of the body are smooth and streamlined; the underbody is wide and flat. Due to the combination of relatively short legs and wide shoulders, if ducks (particularly fast-growing ducklings) are accidentally flipped upside down, it can be difficult for them to right themselves. For this reason, when I am setting up a brooding area, I make sure that there are no significant depressions where a duckling could get trapped on its back.

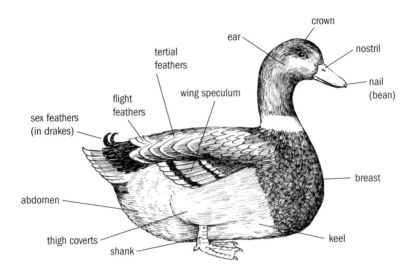

Feathers

Endowed with an extraordinary abundance of feathers, particularly on the underside of their bodies, ducks can swim in the coldest water and remain comfortable. Soft, insulating down feathers close to the skin are covered by larger contour feathers. Several times each day ducks faithfully preen and oil their feathers. A water repellent is produced by a hidden oil gland at the base of the upper tail. As a duck preens, it massages its oil gland with its bill and spreads the excretion onto its feathers. A difference between Muscovies and Mallard-derived ducks is that Muscovies have somewhat less water repellency.

Wings

There are two distinct wing shapes found in domestic ducks. Breeds derived from Mallards have relatively long, narrow, and pointed wings — a good design for fast flight through open airways. Interestingly, in Mallard-derived breeds that are larger than their wild ancestors, wing size did not increase at the same rate as body size during domestication. Therefore, most domestic breeds have lost their ability for sustained flight. However, domestic breeds approximately the size of wild Mallards or smaller have retained their flying skills to varying degrees.

Muscovies have wide, rounded wings designed for slower flight while maneuvering through tropical rainforests. In Muscovies, as body size has increased during domestication, wing size has also increased. Many domestic Muscovy females have retained their flying skills; however, well-fed males from the largest strains are often too heavy for sustained flight.

Tail

Breeds that descend from the Mallard possess short (1½ to 3 inches; 4 to 8 cm), rounded tails; mature drakes have several curled sex feathers in the center of their tails during nuptial plumage. Muscovies have longer, squarish tails that they frequently wag from side to side. Muscovy tails are typically 5 to 7 inches (13 to 18 cm) long and do not have distinctive sex feathers.

Bill

Long and broad, duck bills are well adapted for collecting food from water, catching flying insects, and rooting out underground morsels. With their nostrils located near their heads, ducks can dabble in shallow water and breathe at the same time. Due to hormonal changes in their bodies, female ducks of Mallard ancestry with yellow or orange bills frequently develop dark spots or streaks on their bills when they begin to lay. These blemishes are not a sign of disease. Some waterfowl breeders believe they are an indication of higher than average egg production.

With their long, broad bills, ducks are adept at catching flying insects, straining minute plants and animals from water, and probing in mud. The hooked nail at the end of the bill is used for nipping off tough roots and stems.

Eyes

The vision of ducks is much sharper than our own. Due to the location of their eyes, ducks can see nearly 360 degrees without moving their heads. This feature makes it possible for a feeding duck to keep a constant lookout for danger. Because most of their field of vision is monocular, they do not have excellent depth perception.

Feet and Legs

With webbed toes and shorter legs than land fowl, ducks are not exceptionally fast on land, but they are highly skilled swimmers, preferring to escape danger by way of water. The legs of ducks are relatively fine boned and are more easily injured than those of chickens. Therefore it is advisable to pick them up by their body (for all sizes) or the base of their wings (for small and medium-size birds). When catching ducks, you can also gently grasp them by the neck to gain control, since their necks are relatively strong.

Muscovies have impressive talon-like toenails that enable them to perch securely in trees and to defend themselves. To reduce the chances of sustaining nasty lacerations on my hands and arms, I wear gloves and long sleeves when handling Muscovies.

Back

Normally, a duck's back is moderately long with a slightly convex top line. In Rouens a long back with a noticeable but moderate arch from the shoulders to the tail is a breed characteristic. However, ducks of all breeds with abnormally humped backs should be avoided for breeding purposes since they typically have poor fertility.

Keel

Ducks of Mallard descent that are heavily fed and selected for large size and fast growth often develop keels. A keel is a fold of skin that hangs from the underbody and in extreme cases may run the entire length of the body and brush the ground as the bird walks. A well-developed keel is a breed characteristic of Aylesburies and Rouens but is not preferred on any other breed.

Behavior

Just as you and I have distinct personalities, so each duck has its own peculiarities and habits. However, ducks tend to follow certain behavioral patterns that you should understand if you're going to do a good job of raising them.

Imprinting

It is almost magical — the way a newly incubator-hatched duckling will devotedly follow a human who is many hundreds of times larger than itself. The phenomenon of day-old animals following the first vocal and animated being it encounters has been labeled imprinting. This highly developed behavior, which has been observed in most species of precocious birds and animals, ensures

FAITHFUL FOLLOWERS

To gain a better firsthand understanding of imprinting, I set up a small antique, still-air incubator (no fan or motor that would produce background noise or vibrations) in our living room and filled it with duck eggs. I turned the eggs by hand three times daily, and I made a point of talking to the eggs in a modulated and soothing voice, especially during the last 10 days of incubation. On the 24th day I stopped opening the incubator and turning the eggs; however, I continued to talk to the eggs three times daily. On the 28th day, a carefully orchestrated protocol was followed: Three assistants and I sat on the floor, evenly spaced, 6 feet (about 2 m) apart. One at a time, I had my helpers open the incubator and remove one duckling (during this part, no one talked, and those seated did not move). After the assistant sat down and placed the duckling in the center of the circle, we sat still and all began talking at the same time. Without fail, every duckling headed straight for me! Clearly, since they had never previously seen me, they had imprinted on my voice.

I have tried a similar experiment several times, but using eggs that had been incubated for the first 26 days by ducks, then hatched in an incubator. Not only did these ducklings not respond positively to my voice, but in most cases they were frightened by it.

that babies will not wander off or follow just any critter they happen to encounter during their daily sojourn. In my experience vocal imprinting predates and trumps visual imprinting in ducklings. Observable imprinting behavior is the strongest during the first few days after hatching. Ducklings that are shipped across the country as day-olds can still imprint to a soothing voice upon their arrival.

Not only is imprinting a pleasure to experience, but knowing how to use it is a tremendous aid in raising calm, confident, friendly ducks. As we approach our brooding facilities, we call out in a soothing voice, to let the ducklings know that we are coming — so that when we open the brooder building door, they are not startled by our sudden appearance. Having a handful of chopped, succulent greens further entices them to anticipate and appreciate our arrival.

Even if you are raising only a couple of ducklings in a box in your utility room, it is good to always talk to them in a calm voice well before you are within sight, rather than suddenly appear looming over them — which can cause them to huddle terrified in the corner.

Once ducklings reach maturity, they sometimes fairly abruptly stop tracking the person they have imprinted on and integrate with other ducks.

Pecking Order

This bird law regulates the peaceful coexistence of the duck flock. In simplistic terms, the number-one bird in the flock can dominate all others, the number-two bird can dominate all but number one, the number-three bird can dominate all but numbers one and two, and on down the line.

If you introduce a new bird into an established flock, the existing pecking order is threatened, normally resulting in a power struggle that may evoke fighting. Unless the conflict is causing serious injury to the participants, you should not intervene, remembering that the roughhousing is necessary for the future peace of the flock.

Feeding

The natural diet of Mallards and their derivatives consists of approximately 90 percent vegetable matter (seeds, berries, fruits, nuts, bulbs, roots, and succulent grasses) and 10 percent animal matter (insects, mosquito larvae, snails, slugs, leeches, worms, and an occasional small fish, tadpole, or rodent). Muscovies typically are a bit more carnivorous. Sand and gravel serve as grinding stones in the gizzard.

Ducks feed by dabbling and tipping up in shallow water, drilling in mud, and foraging on land. While ducks eat tender grass and greens, they're not true grazers, as geese are.

Swimming

Skillful and enthusiastic swimmers from the day they hatch, ducks will spend many happy hours each day bathing and frolicking in water if it is available. However, ducks of all types and ages (particularly Muscovies and ducklings) can drown if their feathers become soaked and they are unable to climb out of the water. For this reason ducks must be allowed to swim only where they will be able to exit easily. In general it is safest (unless you are close by) to keep ducklings out of water until they are at least two to four weeks old. Ducks can be raised successfully without swimming water.

Mating

Ducks will pair off, although domesticated drakes often mate with a variety of females in a flock. If you raise several breeds and wish to hatch purebred offspring, each variety needs to be penned separately 2 to 4 weeks prior to and throughout the breeding season.

The number of females that a drake will successfully fertilize is variable, so the following guidelines should be used as suggestions only and modified as needed for any particular situation. In single-male matings, a drake can usually be given two to five females, although males sometimes have favorites and may not mate with the others. In flock matings one vigorous six- to eighteen-month-old male for every four to seven females gives satisfactory fertility in most breeds. For good fertility, females need to be bred by drakes at least once every 4 or 5 days. A few eggs that have been laid as long as 10 days after copulation may be fertile.

For the health and safety of the females, excessive drakes should always be penned separately. Drakes tend to be very competitive. If they are in a pen with or near other drakes, they will often be much more aggressive in their breeding behavior and frequency. Drakes seem to "know" that the last one to mate with a duck will sire the majority of ducklings. If there is only one drake within sight or sound of females, he will often be much more complacent in his breeding behavior. Young drakes, six to twelve months of age, typically are fertile earlier in the breeding season than are older drakes. Aged drakes sometimes will not be fertile until much later in the breeding season, when natural day length and temperatures have increased. Some drakes will remain fertile until they are ten years of age or older, but typically their best breeding age is six months to three years.

Water is an aphrodisiac for ducks. While they typically prefer to mate in water — even if it's a shallow puddle — most breeds can copulate successfully on land. Some large breeds, especially deep-keeled Aylesburies and Rouens, often have higher fertility if they have access to water at least 6 inches (15 cm) deep.

The genitalia of waterfowl is considerably more complicated than that of land fowl, making fertility issues somewhat more prevalent in ducks than in chickens. For this reason I have found it prudent to keep backup males for valuable matings.

Nesting

In the wild a Mallard duck will hide her nest in a protected spot, such as among tall grass or under a bush. The nest is typically a shallow depression in the ground that is excavated with her feet and toenails, and then lined with twigs, grass, leaves, and moss that she carries in her bill from the surrounding area. Normally the duck stealthily returns to the nest once daily to lay an egg until a clutch of 8 to 15 eggs is reached. As the nest fills with eggs, the duck will pluck feathers and down from her underbody to add insulation to the nest and to make a brood patch on her body so that the eggs will be in closer contact with her warm skin once incubation commences. During the incubation period, the duck sits tightly on the nest and leaves only for a short daily excursion to defecate, feed, drink, and bathe.

To minimize egg losses to predators (including egg-loving crows and jays) and yield clean eggs that are suitable for incubation or human consumption, provide ducks with a safe nesting area that has an ample supply of dry, clean nesting material such as straw, hay, or wood shavings. Some ducks, especially those from strains that have been selected for high egg yields, will lay their eggs at random on the ground or even while swimming.

Muscovies will use ground-level nest boxes, but if given a chance, they often prefer elevated nesting sites such as tree hollows, hay lofts, or raised nest boxes.

Fighting

In established flocks that are not overcrowded and have a proper sex ratio, ducks get along well together and rarely fight. If a new bird is introduced into a flock, there will be a short period of chasing, pushing, and wing slapping, but normally such conflicts subside without human intervention. Adult ducks seldom inflict serious injury on one another unless there are an excessive number of drakes in a flock or they are overcrowded. As a rule small groups of two to five drakes will not fight among themselves if there are no females around. Fighting tends to be more prevalent as testosterone levels increase due to long or increasing natural or artificial day length. One way to reduce fighting in groups of drakes is to "light neuter" them by keeping them in a totally dark enclosure for 14 to 18 hours a day.

Herding

With practice, one person can fairly easily move a group of ducks from one location to another by herding them. Unlike chickens and Muscovies—which

HERDING DUCKS AFTER DARK: THE LIGHT WAND

To protect ducks from nocturnal predators such as raccoons, foxes, bobcats, coyotes, skunks, and owls, they should be locked in a predator-proof enclosure every night. In many localities, the safest strategy is to put them in at least half an hour prior to darkness. If you are tardy putting them in their nighttime quarters, they can range a fair distance from home, since they are enthusiastic foragers of earthworms and other delicacies that emerge at nightfall.

When I need to herd ducks after nightfall, I use a good flashlight with a strong beam as a light wand to herd them back to their enclosure. By shining the light several feet behind the ducks and slowly sweeping the beam back and forth, I guide them back to their quarters. Do not shine the light directly at them, as this can cause them to panic and scatter. With the right flashlight, good technique, and a little practice, you can gather a scattered flock of foraging ducks from several acres in just a few minutes.

tend to scatter in many directions — ducks of Mallard descent have a strong inclination to form a tight group when they feel pressure to move in a given direction by a human or other predator. This is why they are used extensively for training herding dogs. With arms outstretched the wise duck keeper can use this trait to minimize the need to catch ducks and simply walk them to a desired location. For large flocks we sometimes use a long bamboo pole held horizontally in front of us to give greater control.

Life Expectancy

Ducks can live a surprisingly long time when protected from accidental deaths. It is not unusual for ducks to live and reproduce for six to eight years, and there are reports of exceptional birds living fifteen years and longer. However, few ducks die of old age (except for those special pets!), since fertility and egg production often decrease after three to five years; hence, many owners do not consider it economical to keep ducks past this age.

HOUDINI DUCKLINGS

Belknap Creek meandered through pastures and marshes near my childhood home. As I explored its marvels, one of my favorite activities was hiding in the tall grasses along the banks, watching wild ducks with their broods.

One day, as a Mallard and her seven newly hatched ducklings swam by, I couldn't resist jumping into the water and trying to catch them. As soon as I hit the water, the mother feigned a wing injury in an attempt to lure me away from her brood. The ducklings vanished. The muddied water soon cleared as I stood knee-deep in the creek, and to my surprise I saw several of the ducklings clinging to underwater vegetation, attempting to hide from me. I quickly climbed out of the water and concealed myself in the tall grass again. Within a few seconds the ducklings bobbed to the surface and swam off under the watchful eye of their mother.

3

Choosing the Right Duck

CHOOSING AN APPROPRIATE BREED can be enjoyable and plays an important role in determining the success or failure of your duck project. Unfortunately, novices often assume that a duck is a duck and acquire the first quacking, web-footed fowl they find. This mistake frequently results in expensive eggs or meat, needless problems, and a discouraged duck keeper. Investing a little time at the outset in acquainting yourself with the basic characteristics, attributes, and weaknesses of the various breeds will go a long way toward eliminating unnecessary disappointments.

Important Considerations

The following questions are designed to help you identify your own special requirements when choosing a breed.

Purpose

What is your main reason for raising ducks? Is it for pets, eggs, meat, feathers, decoration, exhibition, insect and slug eradication, aquatic plant and algae control in ponds, or a combination of these and other aims?

Location

Where are you located? Some breeds are noisier than others—a fact that should be taken into consideration when you have neighbors in close proximity. Talkative breeds (Calls and Pekins in particular) also attract more predators.

Climate

What are your temperature extremes? All breeds are adaptable to an impressive range of climates. However, when the thermometer falls below 15°F (–9°C), the bantam breeds (especially Call and East Indie), the slender and tight-feathered Runner, and the bare-faced Muscovy can benefit from tighter housing than other breeds (see chapter 14 for more on housing).

Management

How are you going to manage your ducks? Will they be locked up in a building, enclosed in a fenced yard, put on a pond, or allowed to roam freely? Will you provide all or most of their food or only a small supplement to what they glean for themselves? Certain breeds adapt better than others to total confinement (especially Call, East Indie, Miniature Appleyard, Silkie, Australian Spotted, Muscovy, and Pekin). Some breeds stay close to home while others will roam over a large area. Foraging ability also varies considerably. And due to small size or a trusting temperament, some breeds are more likely to be taken by predators.

Color

What plumage color is best suited to your situation? Color is significant for several practical reasons. The pinfeathers of light-plumaged birds are not as visible as those of birds with dark plumage, making it easier to obtain an attractive carcass with light-colored ducks. However, in many circumstances, medium to dark birds are better camouflaged, making them less susceptible to predators. Also, if there is no bathing water available, colored ducks generally maintain a neater appearance than do white ones.

Preference

What do you like? Raising ducks should be enjoyable, so choose a breed that you find attractive and interesting. Read the following chapters on the various breeds, look at their pictures, and choose one or more breeds that fit your tastes and needs. For most people and situations, there are several breeds that will work well. Some folks raise one breed to produce only eating eggs, another primarily for meat, and a third for natural incubation or decoration.

I'm frequently asked what my favorite breeds are. After working with every recognized breed and many that are not officially recognized, I can honestly say that I like them all.

Population Status

Do you wish to help save a rare breed? The population status of different breeds ranges from numerous to endangered. When you purchase and raise a rare breed, you're helping improve its chances of continued survival.

Useful Terminology

In order to read and speak intelligently about the different breeds of ducks, it is useful to understand some basic terminology.

- **Species.** The domestic ducks raised in North America belong to one of two species. Those that are descendants of the common wild Mallard go by the scientific name *Anas platyrhynchos domesticus*. The one breed that descended from the wild Muscovy is *Cairina moschata domesticus*. (No, you don't need to memorize these scientific names, unless you are working on an advanced degree in Quackology!)
- **Class.** The American Poultry Association has divided domestic ducks into four classes, based on size: Bantam, Lightweight, Mediumweight, and Heavyweight. When ducks are entered in most shows, their classes must be identified on the entry forms.
- **Breed.** Over the years waterfowl breeders in many parts of the world have developed distinctive types of ducks. As ducks in particular regions became uniform in physical characteristics, they were given names and recognized as separate breeds. Those breeds that have been effectively promoted and proven to have desirable traits have survived. A recognized breed is one that has been officially accepted into the standard of one or more of the governing organizations of purebred poultry.

BODY PROFILES OF BREED TYPES

Campbell   Indian Runner   Rouen   Call

- **Breed type.** Most breeds can be identified by their unique type. Type refers to the shape and dimensions of the head, neck, and body and the way these parts all fit together to form the overall silhouette that is characteristic of a breed. A bird is said to be "typey" if it possesses all of its breed's key characteristics for shape. (Some breeds, such as the Cayuga, Orpington, and Swedish, have similar size and type and therefore are distinguished primarily by the color and pattern of their plumage.)
- **Variety.** Some breeds have more than one variety. In ducks varieties almost always are distinguished from one another by the color or pattern of their plumage. An example of a breed that has just one variety is the White Aylesbury. On the other hand, Runners come in eight recognized varieties: White, Black, Blue, Chocolate, Penciled, Fawn & White, Gray, and Buff. A recognized variety is one that has been officially accepted into the standard of one or more of the governing organizations of purebred poultry.

  Some breeds have unrecognized varieties that are created by hobbyists or commercial producers. These can become recognized if they go through the standardization procedure required by the American Poultry Association or the American Bantam Association (see chapter 8, page 100).
- **Strain.** Breeds consist of different strains. A strain is a group of birds that all descend from one flock or breeding farm and are more closely related than the members of the breed at large. Strains are usually identified by using the name of their originator as a prefix; for example: Horton East Indies, Lundgren White Calls, Oakes Khaki Campbells, Sherraw Rouens, or Legarth Pekins.

  When acquiring ducks, keep in mind that the strain is at least as important as the breed. For example, there are strains of Khaki Campbells that will consistently produce 300 to 340 eggs per female per year when managed properly. Conversely, there are other strains of Khaki Campbells that will barely lay 150 eggs per female per year. A broader example is the difference between production-bred and standard-bred strains as described in chapter 11.
- **American Poultry Association.** The APA is a membership organization that sanctions shows and publishes standards for the breeds and varieties it recognizes in chickens, ducks, geese, and turkeys.
- **American Bantam Association.** The ABA is a membership organization that sanctions shows and publishes standards for the breeds and

varieties it recognizes in bantam chickens and ducks. The ABA and APA standards do not always agree. See appendix H, page 341, for ABA contact information.

- ***American Standard of Perfection.*** This book contains a complete description of the ideal specimen in each breed and variety recognized by the APA. The *Standard* is periodically revised and updated and is available from the APA (see appendix H, page 341, for address). Most serious breeders and exhibitors own a copy. Local poultry clubs and public libraries sometimes have one available for loan.
- ***Bantam Standard.*** This book contains a complete description of the ideal specimen in each breed and variety recognized by the ABA. The ABA publishes standards only for bantam breeds, and their standards have minor differences from the APA's standards.
- **Faults.** In purebred poultry a fault is any characteristic of a bird that falls short of the ideal as described in the *Standard of Perfection*. It is frequently said that the perfect specimen has never been raised, and even the best birds have at least minor faults. Even among the most experienced and respected breeders and judges, complete agreement on the severity of faults does not often occur. What one person identifies as a significant fault, another may overlook.

Breeds of Ducks

Most domestic ducks raised in North America belong to one of 23 breeds or their hybrids. These water-loving fowl represent a marvelous cornucopia of sizes, shapes, and colors. By examining their breeding histories, we get an intriguing glimpse of the places, events, and traditions that have helped to shape the present duck world.

When reviewing the various breeds, always remember that birds within the same breed can vary greatly in their physical, practical, and personality traits. Furthermore, the environment they are raised in and the diet they consume can significantly alter not only their growth and productivity but also their appearance. For detailed descriptions of plumage colors and patterns, see chapter 10, Understanding Duck Colors.

BREED PROFILES

Weight Class	Breed	Male/Female Pounds (kg)		Yearly Egg Production
BANTAM	Australian Spotted	2.2 (1)	2 (0.9)	50–125
	Call	1.6 (0.7)	1.4 (0.6)	25–75
	East Indie	1.8 (0.8)	1.5 (0.7)	25–75
	Mallard	2.5 (1.1)	2.2 (1)	25–100
	Mini Silver Appleyard	2.2 (1)	2 (0.9)	50–125
	Silkie	2.2 (1)	2.0 (0.9)	50–125
LIGHT	Bali	5 (2.3)	4.5 (2)	120–250
	Campbell	4.5 (2)	4 (1.8)	250–340
	Harlequin	5.5 (2.5)	5 (2.3)	240–330
	Hook Bill	4 (1.8)	3.5 (1.6)	100–225
	Magpie	6 (2.7)	5.5* (2.5)	220–290
	Runner	4.5 (2)	4 (1.8)	150–300
MEDIUM	Ancona	6.5 (3)	6 (2.7)	210–280
	Cayuga	8 (3.6)	7 (3.2)	100–150
	Crested	7 (3.2)	6 (2.7)	100–150
	Orpington	8 (3.6)	7 (3.2)	150–220
	Swedish	8 (3.6)	7 (3.2)	100–150
HEAVY	Appleyard	9 (4.1)	8 (3.6)	200–270
	Aylesbury	10 (4.5)	9 (4.1)	35–125
	Muscovy	12 (5.4)	7 (3.2)	50–125
	Pekin	10 (4.5)	9 (4.1)	125–225
	Rouen	10 (4.5)	9 (4.1)	35–125
	Saxony	9 (4.1)	8 (3.6)	190–240

Note: Information presented in this profile is based on the average characteristics of each breed. Actual performance may vary considerably from the norm.

*The APA Standard gives weights of 5 pounds (2.25 kg) and 4.5 pounds (2 kg), respectively, for Magpie drakes and ducks, approximately 1 pound lower than typical.

Egg Size per Dozen Ounces (kg)	Mothering Ability	Foraging Ability	Status
20–24 (0.55–0.7)	Excellent	Excellent	Endangered
16–20 (0.45–0.55)	Fair–Excellent	Poor–Fair	Common
18–24 (0.5–0.7)	Excellent	Excellent	Fairly common
24–28 (0.7–0.8)	Excellent	Excellent	Abundant
20–24 (0.55–0.7)	Excellent	Excellent	Endangered
20–28 (0.55–0.8)	Excellent	Excellent	Endangered
28–36 (0.8–1.0)	Poor–Fair	Excellent	Endangered
28–34 (0.8–0.95)	Poor–Fair	Excellent	Fairly common
29–34 (0.8–0.95)	Poor–Good	Excellent	Rare
24–32 (0.7–0.9)	Fair-Good	Excellent	Endangered
30–38 (0.85–1.1)	Fair–Good	Excellent	Rare
28–36 (0.8–1.0)	Poor–Fair	Excellent	Common
30–38 (0.85–1.1)	Fair–Good	Excellent	Endangered
30–38 (0.85–1.1)	Fair–Good	Good	Common
30–38 (0.85–1.1)	Fair–Good	Good	Common
30–36 (0.85–1.0)	Fair–Good	Good	Fairly common
30–38 (0.85–1.1)	Fair–Good	Good	Fairly common
34–40 (0.95–1.1)	Fair–Good	Good	Rare
38–44 (1.1–1.25)	Poor–Fair	Fair	Rare
38–50 (1.1–1.4)	Fair–Excellent	Excellent	Abundant
36–46 (1.0–1.3)	Poor–Fair	Fair	Abundant
36–44 (1.0–1.25)	Poor–Good	Fair–Good	Common
36–46 (1.0–1.3)	Fair–Good	Good	Rare

4

Bantam Breeds

THE BANTAM CLASS INCLUDES THE MINIATURE BREEDS of the domestic duck clan. These birds weigh between 18 and 40 ounces (0.5 and 1 kg). They are popular for pets, decoration, and exhibition but also are good seasonal layers and produce meat that has outstanding flavor and fine texture. Highly adaptable, they thrive in limited space under close confinement or can be allowed to roam at large where predators are not a problem.

Under many conditions, a patrol of 8 to 20 of these little broad-bills per acre (0.5 ha) of land provides an effective means of controlling many pests, including mosquito larvae, various insects, snails, and small- to medium-size slugs (large slugs are best dealt with by larger ducks). During the seasons of the year when there is an abundant supply of natural foods, bantam ducks that roam over a large area during the day require minimal supplemental feed.

Bantam-Raising Hints

Basic care of bantam ducks is similar to that of their larger relatives. However, due to their diminutive size, keep the following in mind.

Keeping Bantams Safe

Because of their size, bantam ducks are more susceptible to winged predators (hawks during daylight hours and at night large owls) than are bigger ducks. For this reason in some localities it is prudent to pen bantam ducks in a covered enclosure. If left out at night on a pond, they have a better chance of surviving if there is an island with a covered area where they can hide from owls.

Most bantam ducks are capable aviators, but in 50 years I have not had homegrown birds depart through the airways. However, do not expect them to stay at home if they are terrorized by predators (including dogs and people), are frequently allowed to go hungry, or are overcrowded. When relocating them to a new home, it is safest to put bantams in a tightly fenced pen with a covered top until they are acclimated to their new residence. They can also be grounded by clipping the primary flight feathers of one wing.

Size Considerations

Diminutive size is primarily genetically controlled. Bantams tend to produce some offspring that are larger than their parents. Therefore, to maintain the correct size, it is important to select small birds in each generation for breeding purposes. A common technique for enhancing the production of small ducks is to use linebred matings. This entails mating together related birds, such as first cousins, uncles, or aunts to nephews or nieces, or parents to offspring. If vigor and productivity decline due to excessive inbreeding, a carefully selected outcross (the mating together of birds that are not closely related) to another family or strain will usually correct the problem. Typically, when you mate birds that are more distantly related, their offspring will be larger in size and more productive.

Nutrition and environment also affect the body size of bantam ducks. Ducklings that are grown out on a 15 percent protein ration that is balanced for all essential nutrients typically will be smaller than those provided a high-protein diet. A feed regimen employed by some successful bantam breeders consists of giving ducklings chopped leafy greens (such as dandelions, leaf lettuce, or succulent young grass that has not been contaminated with chemicals) and chick-size insoluble granite grit and feed made up of one part (by volume) uncooked rolled oats to two parts good-quality 18 to 20 percent protein waterfowl starter/grower crumbles. Ducklings that get out at an early age to forage and exercise extensively will generally not grow as large as those that are raised in close confinement and do little more than eat, drink, and sleep.

Breeding

To maintain productivity from generation to generation, breed only from females that lay well-shaped eggs with good shells. To encourage good production and high hatchability of eggs, it is helpful to supply bantam breeding ducks with a high-quality waterfowl breeder or game-bird breeder ration containing approximately 20 percent protein during the breeding season (2 to 3 weeks prior to the first eggs and throughout the laying period).

In single-male matings, pairs or trios often produce the best fertility, although some drakes will successfully fertilize three or more ducks. In multiple male flock matings, it is often necessary to have at least two to three females per male to minimize fighting among the drakes. Most females are excellent natural mothers and sometimes are used for hatching the eggs of endangered wild ducks. Here at the Waterfowl Preservation Center, we have employed bantam females with excellent success to hatch eggs from lines of birds that hatch better under natural incubation.

Australian Spotted

The charming and beautiful Australian Spotted, its "down under" name notwithstanding, originated in the United States. The late John C. Kriner Jr., of Orefield, Pennsylvania, told me in communications during the mid-1970s that he and Stanley Mason developed this breed (originally known as the Australian Spots) in the 1920s and began exhibiting them in 1928. Kriner stated that the foundation stock was the Mallard, Call, Northern Pintail, and an unidentified wild Australian duck possessing spotted plumage.

The original birds were kept together and allowed to interbreed for several generations. The preferred offspring were then selected and bred from, thus forming the new breed. Because Pintail × Mallard crosses usually are sterile, I originally assumed that Pintails had not contributed to further generations. However, after raising and studying Australian Spotted for two decades and observing physical and behavioral characteristics normally not seen in pure Mallard derivatives, I am not as ready to dismiss Pintails as one of their ancestors. Every once in a while, a mule produced by a donkey × horse mating is fertile. Likewise, it is possible that Kriner and Mason had a rare fertile Pintail × Mallard hybrid that did contribute genetic material to their new breed.

During a 1975 visit by me to the late Henry K. Miller's Blue Stream Farm, located near Lebanon, Pennsylvania, Mr. Miller showed me his birds and told me about developing his own strain of Australian Spotted in the 1940s. Originally, this breed was classified and exhibited as a wild species.

Prior to 1990 this exquisite little duck was not readily available to the public. Thankfully, its status today is improved, although the number of Australian Spotteds is still dangerously low.

This six-month-old Greenhead Australian Spotted drake, bred on our farm, is in full nuptial plumage. He has the desired conformation and plumage pattern. Drakes in juvenile or eclipse plumage have bold spotting on their bodies.

A six-month-old Greenhead Australian Spotted duck, bred on our farm, possessing excellent conformation and boldly spotted plumage pattern.

Description. According to its originators, the Australian Spotted should be intermediate in type between the racy, long-billed Pintail and Mallard and the plump, short-billed Call. The bill is of medium length (1¾ to 2 inches [4.5 to 5 cm] long) and medium width. The head is oval, of medium size, and moderately streamlined, without the distinctive high forehead and puffy cheeks of the Call duck. The body is moderately racy and has a teardrop-shaped profile when viewed from both the side and top. The legs are attached near the center of the body, allowing for a nearly horizontal body carriage when the bird is relaxed. The Australian Spotted is a bit smaller than most wild Mallards, at 30 to 38 ounces (0.9 to 1 kg).

The vocalizations of the Australian Spotted are not as high-pitched or plentiful as those of Calls. Here at the Waterfowl Preservation Center, Australian Spotteds are one of the favorites due to their friendly, inquisitive, and enthusiastic natures.

Varieties. Greenhead is the original color, with Bluehead and Silverhead varieties having been developed here at the Waterfowl Preservation Center. The variety name refers to the color of the drake's head when in breeding plumage. (See chapter 10 for color descriptions and information.)

Selecting breeders. Choose active, bright-eyed birds that have good breed type and moderately small size. Australian Spotted should not be as tiny as Calls or East Indies. Due to the wide genetic base that this breed originated from, there is considerable variability in the plumage patterns expressed by each generation. For this reason it is important to select breeding birds that will produce the highest percentage of appropriately colored offspring. In particular I look for drakes that have bold spotting on the sides of their body and over the shoulders when in juvenile and eclipse plumage. In their nuptial plumage they should have a white ring that encircles the neck, and the chestnut-burgundy color of the breast should extend strongly onto the shoulders and sides of the body. In females look for bold spotting. (For more details, see chapter 10.)

Selecting and preparing show birds. At this writing, the Australian Spotted is not yet in the APA's *Standard of Perfection*. They can be shown at most poultry shows, but they cannot compete against standardized breeds. In size and conformation a good Australian Spotted looks as though it could be a wild duck. Avoid birds that have strong Call characteristics. Australian Spotteds are tidy birds, and if you provide them with a balanced diet, a clean pen with sufficient room, fresh bathing water, and plenty of shade, they will prepare themselves for shows.

During the molt and for 6 to 8 weeks prior to a show, use the ration recommended for Calls and Mallards to encourage excellent condition.

Comments. Along with their diminutive size and delightful plumage, Australian Spotteds have proven to be personable and long-lived. They are exceptionally hardy and are outstanding foragers, and the females — along with those of the Miniature Silver Appleyards — are the best layers of the Bantam breeds. Their cream, blue, or green eggs hatch into lively ducklings that display a tremendous zest for life. They are the quickest-maturing breed I have observed. Sometimes, three- to four-week-old drakelets engage in courtship displays that I have not observed in other breeds until at least twice that age. We have had ducks commence laying at just over three months of age.

Because of the rarity of the Australian Spotted, their future is tenuous. They merit being saved, not only for their wonderful aesthetics and personality but also for their many practical attributes.

Call

In the twenty-first century, this miniduck is primarily raised for exhibition, decoration, and pets. The current pampered status of the Call is in stark contrast to its humble beginnings as a working, turncoat duck in Western Europe.

A descendant of the wild Mallard, the Call probably originated in Holland, where it still goes by the name of Decoy Duck. Originally employed as live decoys, Calls were used to entice wild ducks to enter large funnel traps. Later, market hunters tethered Calls near their gunning stations to lure wild ducks to fly within shooting range. The Call may be the only breed of duck in history to have been originally selected for its voice. Their unique high-pitched voices (which carry over long distances) and ability to talk fast and with great persuasion made them invaluable sidekicks to generations of hunters.

In the early decades of its development, the physical appearance of the Call was hardly distinguishable from its wild ancestors. Over the years the bill was shortened and the body size reduced through selective breeding. Both the original Gray variety and the White sport were included in the first *Standard of Perfection* published by the American Poultry Association (APA) in 1874.

Since the middle of the twentieth century, Calls have made impressive strides in both popularity and quality. In North America today this little charmer is a favorite among hobbyists. At poultry shows Calls win more duck championships than any other breed.

Description. Modern Calls look like toy ducks. A tiny, plump, bowl-shaped body; short, broad bill not much larger than a thumbnail; stubby neck; large, round head that is wide across the skull; and short legs positioned near the center of the body are desired characteristics. Along with being the most talkative breed, they are also one of the most active. They do everything with vim and vigor — flirting, searching for tasty tidbits, bathing, preening, and flirting some more. Most Calls are easily tamed and if talked to frequently will carry on animated conversations with their human caregivers.

Varieties. Calls are raised in virtually every color and pattern known in domestic ducks, with new varieties frequently coming on the scene. Currently, standard colors include Gray, White, White-Bibbed Blue, Butterscotch, Snowy, Buff, Pastel, Blue & White Magpie, and Black & White Magpie. Nonstandard colors include, but are not limited to, Saxony, Aleutian, Spot, Self-Blue, Self-Chocolate, Self-Black, White-Bibbed Black, White-Bibbed Chocolate, Ancona, Penciled, Fawn & White, Khaki, Cinnamon, Dusky, Blue Dusky, Lilac, Blue Fawn, Yellow-Belly, Pied, and Crested. (See chapter 10 for color descriptions and information.)

A White Call old drake with the plump body, short neck, large round head, chubby chin, full cheeks, short bill, and horizontal carriage desired in this breed. A multiple winner bred by Gary and Kari Bennett of Oregon.

Selecting breeders. When a breed is selected for extremely small size, it is especially important to choose breeding stock that is vigorous, active, strong-legged, and bright-eyed. Beginners often make the mistake of spending top dollar to buy extremely small Calls with the shortest bills they can find, only to discover that these specimens may be fine show birds but are poor or nonproducers.

As a rule your best bet is to select a small, typey male that displays abundant vigor and mate him with one or more females that have excellent type and heads but are a bit larger in size so they have the capacity to produce viable eggs. If they are descendants of a strain that produces ducks with short bills, you should know that breeding birds do not need to have extremely short bills themselves to produce offspring with excellent bills.

Selecting and preparing show birds. The mantra of the Call exhibitor is, "Everything else being equal, the smaller the size of the bird and the shorter the bill, the better." Most of today's top show specimens have bills measuring 1 inch (2.5 cm) or less in length. Also important are large, round heads with high crowns that rise abruptly from the base of the bill, full cheeks, bodies that are short and wide, horizontal body carriage, and good color.

Calls normally are such tidy birds that when provided with a balanced diet, clean pen, fresh bathing water, and plenty of shade they will just about prepare themselves for shows. During the molt and for 6 to 8 weeks prior to a show, a ration that encourages excellent feather condition can be made by mixing six parts (by volume) game-bird flight conditioner, three parts uncooked oatmeal or oat pellets, and one part cat kibbles. Chick-size granite grit should be supplied free choice and greens such as lettuce or chard supplied three or more times weekly. If these greens are put in water, the birds will spend many happy hours dabbling, which encourage them to keep their heads and bills sparkling clean.

Comments. Calls have not been used extensively as decoys for nearly a century, but their loud, persistent talking is a reminder of their bygone vocation. Some people are charmed by the chatter; others find it annoying. When deciding where to locate your Call pen, it is wise to consider the proximity to your bedroom or any area you wish to remain quiet! Call eggs are usually white or tinted green or blue. Day-old Calls are often more delicate than the ducklings of other breeds; special care should be taken to ensure that they are sufficiently warm and do not get wet. Once they are well started, Calls are typically quite hardy.

Of all breeds of domestic ducks, Calls are the most susceptible to sinus infections, which some people attribute to their short bills and unique head structure. I have found that this problem can be virtually eliminated by being extra vigilant in keeping Calls of all ages in a clean environment and changing their drinking water at least once daily.

East Indie

The "little duck with many names" is an apt label for this widely admired breed. Brazilian, Buenos Aires, East Indian, East India, Emerald, and Labrador Duck are some of the more common names that have been or are still being used in some countries. In North America, however, its official name is East Indie.

Scant information is available concerning its formative years. Even though it is easy to imagine these emerald beauties at home on streams deep in tropical forests, their exotic names apparently have no correlation to their place of origin. It is generally accepted that the East Indie originated in North America during the first half of the 1800s, was then improved by British breeders (who still use the name East Indian) during the last half of that century, and was perfected to its current level of development by North American breeders during the last half of the 1900s.

Many writers have suggested that the East Indie is a descendant of the wild American Black Duck (*Anas rubripes*), which inhabits much of the eastern portion of North America. I have found no hard evidence to substantiate this theory. American Black Ducks are black in name only; their plumage is very dark brown with most feathers edged with lighter brown. The true black plumage of the East Indie is caused by a black mutation that is fairly common in Mallard derivatives. Furthermore, *Anas rubripes* drakes do not have curled sex feathers in their tails, whereas Mallards and their descendants, including East Indies, have this diagnostic marker.

Some of the early writers referred to the East Indie as a smaller and better-colored version of the Cayuga. In his book *The Practical Poultry Keeper* (1886), L. Wright wrote that the East Indian ". . . should be bred for exhibition as small as possible, never exceeding five and four pounds." As recently as the 1960s, 3-pound (1.5 kg) East Indies were common, though generally not preferred. Currently, show-winning birds that I have weighed commonly tip the scales at 16 to 20 ounces (0.5 to 0.6 kg).

The East Indie was included in the first British *Book of Standards* in 1865, as well as in the first APA *Standard of Perfection* in 1874. Today among hob-

A young pair of Black East Indies. The drake won Reserve Champion Bantam
Duck at the 1999 Northwest Winter Classic. Bred by Art Lundgren of New York;
shown by Gene Bunting of Oregon.

byists this is the second most popular bantam duck breed, behind the Call.
Currently, East Indies are raised primarily for decoration, pets, exhibition,
and control of mosquitoes and smaller garden pests.

Description. Good East Indies are intermediate in type, positioned between
the short, plump Call and the long, racy Mallard. Breed characteristics include
a moderately small oval head, bold eyes, moderately long neck of medium
diameter, medium-wide body that is moderately long, medium-length legs,
and horizontal body carriage. The bill has a slightly concave top line when
viewed from the side. Winning birds I have measured had average bill lengths
of 1¾ inches (4.5 cm) in drakes and 1⅝ inches (4 cm) in ducks.

The iridescent green-black plumage of the Black East Indie must be seen
before its beauty can be fully appreciated. When raising black ducks, keep
in mind that those with a high degree of iridescence almost always have at
least a few small white feathers in their plumage — even out of the very best
breeding stock that has been carefully selected for generations. Furthermore,
as they age — especially females with outstanding iridescence — ducks will
develop increasing amounts of white in their plumage. Experienced breeders
know that some females that were winning show specimens as young birds
will eventually have 90 percent or more of their plumage turn white.

Varieties. Black with green iridescence is the traditional color. Bill Mayer of Michigan developed Blue East Indies with fine type and small size and introduced them to the public in 1999 at the Western Waterfowl Exposition in Albany, Oregon. (See chapter 10 for color descriptions and information).

Selecting breeders. Choose active, bright-eyed birds that have good breed type and horizontal carriage, with legs, feet, and bills as dark as possible (keep in mind that the leg and bill color naturally lighten somewhat during the breeding season and as birds age). As with Calls, the tiniest birds are sometimes not productive, so use the same strategy of choosing a vigorous small male to be mated with females that have enough body capacity to lay viable eggs. (For in-depth guidelines for breeding black plumage, see chapter 10, Understanding Duck Colors.)

Selecting and preparing show birds. To be successful in good competition with East Indies, you must show birds that are small, have excellent type and color, show no white in their plumage (one small white feather can be the difference between a bird's winning first or not placing at all), and are in first-class feather. Because brilliant black ducks normally have at least a bit of white somewhere in their plumage, it pays to go over a bird with a tweezers to remove any small white or faded feathers prior to taking them to a show. It is not cheating to remove white feathers as long as they are not the large wing or tail feathers. However, artificially coloring feathers is not allowed. Some exhibitors wipe down their entrants with a silk cloth just prior to judging to enhance the feathers' sheen.

To improve the green sheen of the plumage, during the molt and for 6 to 8 weeks prior to a show, you can provide a ration consisting of six parts (by volume) game-bird flight conditioner, three parts uncooked oatmeal or oat pellets, one part cat kibbles, and one-half part black-oil sunflower seeds. Chick-size granite grit should be supplied free choice and greens such as lettuce or chard given at least three times weekly.

Comments. East Indies normally are not as talkative as Calls. They can make great pets but typically are a bit shier than Calls. The first eggs of a laying cycle are often gray or black, gradually changing to lighter hues of gray, blue, or green, as the season progresses. In my experience, East Indies are good natural mothers. If not overly inbred, the ducklings are hardy, especially after they are one to two weeks old.

Mallard

The wild Greenhead is native to most countries of the Northern Hemisphere and goes by the scientific name *Anas platyrhynchos platyrhynchos* (a-nas pla-tiring-kos). This much-admired species is believed to be the primary parent stock of all domestic duck breeds except the Muscovy. No one knows for sure when Mallards were first domesticated, but there is evidence that Southeast Asians and Romans were raising ducks in captivity prior to 500 BC.

Mallards are widely raised, primarily for hunting clubs, gourmet meat, decoration, pets, and exhibition. Interestingly, the APA did not include them in their *Standard of Perfection* until 1961.

Description. Authentic Mallards are elegantly streamlined. They have trim, teardrop-shaped bodies, slender necks that can telescope to surprising lengths, aerodynamic heads, and horizontal body carriage. A distinctive characteristic of the true Mallard is a long, slender bill. In good exhibition drakes the bill measures 2¾ to 3 inches (7 to 7.5 cm) long, whereas those of ducks typically are 2¼ to 2½ inches (5.75 to 6.5 cm) long.

Most of the Mallards sold by hatcheries have been selected for larger body size and higher productivity. In the process, they have lost some of the elegance and refinement of their more authentic cousins.

This authentic Mallard duck, bred on our farm, displays her teardrop-shaped body, wing speculum, and intricately penciled body plumage.

Varieties. The Gray is the original wild color. Rare varieties include the Snowy, White, Golden, Pastel, and Blue Fawn. (See chapter 10 for color descriptions and information.)

Selecting breeders. Mallards readily adapt to domestication. In just a few generations of heavy feeding in captivity, they can increase in size and become less streamlined in type. If you wish to preserve their sleek characteristics, beautiful color, and outstanding foraging ability, it is necessary to select breeding stock in each generation that most closely resembles the wild birds.

Selecting and preparing show birds. Small, streamlined Mallards with long bills and proper color compete successfully with the best of other breeds. In drakes look for distinct white neck collars and dark chestnut breasts with a minimum of foreign color. In ducks select medium-colored birds with distinctly penciled feathers. Mallards are naturally shy and normally require more cage training than other bantam breeds.

During the molt and for 6 to 8 weeks prior to a show, you can use the same ration as recommended for Call ducks to encourage excellent condition.

Comments. Along with the Muscovy, the Australian Spotted, and the Miniature Silver Appleyard, the genuine Mallard is the most self-reliant of all the breeds I have worked with. In the right environment females are outstanding mothers and will hatch a high percentage of their buff, green, or blue eggs. Mallard hens can be valuable for incubating the eggs of wild species and domestic breeds that may be more difficult to hatch in incubators.

In Canada a migratory waterfowl permit is required to possess Mallards. In the United States, a permit is required to capture, gather eggs of, possess, or sell wild Mallards. In most states a permit is not currently required to raise and sell domestic Mallards. However, regulations change from time to time, so you should check with your local Fish and Wildlife bureau for current regulations. Mallards are not allowed in Hawaii at this time, because it is feared they will crossbreed with endangered wild ducks.

Silkie Duck

These web-footed pixies, which look as though they could have stepped out of the pages of a children's fantasy book, owe their existence to the superb duck breeder C. Darrel Sheraw of Pennsylvania. According to Mr. Sheraw, his original Silkies were a surprise product of his normal-feathered White and Snowy Mallards in the 1980s. Mr. Sheraw is one of North America's most skilled, observant, and veteran waterfowl breeders, and he recognized that these uniquely feathered ducks were worth propagating. He established a breeding flock and named them Silkies. Today, these unique ducks are being raised by hobbyists throughout North America.

Description. The most distinctive characteristic of this friendly little duck is its unusual plumage. Silkies' feathers have a soft, somewhat furlike quality to them, without the extreme fluffiness of Silkie chickens. The plumage of their flanks hangs down to form pantaloons. When held in the hand, Silkies are extremely soft to the touch.

A look at their feathers under a microscope reveals that Silkie feathers lack most of the hooklike barbicels of normal feathers. The barbicels function like Velcro by hooking together the thousands of minute barbs and barbules that make up the flat surface of regular feathers. Consequently, most Silkies are incapable of sustained flight.

Silkies are smaller than wild Mallards, typically weighing between 30 and 38 ounces (0.9 and 1 kg). They have medium to medium-long bills, oval-shaped heads, medium-length legs, and conformation similar to wild Mallards. When relaxed, they stand nearly horizontal.

Varieties. White was the original variety. It is common for some White Silkie ducklings to have some black or gray spotting in their down. These dark markings disappear in the adult plumage in many individuals, although some retain these splashes of color for life. Gray, Dusky, Snowy, and Black varieties have been added to the Silkie family. Due to the unique texture of their plumage, multicolored varieties, such as the Gray, have less distinct borders between color zones than normal-feathered breeds. Also, the under-color shows through, making plumage colors appear softer and more muted in the Silkie. (See chapter 10 for color descriptions and information.)

Selecting breeders. Choose active, bright-eyed birds that are moderately small and have an abundance of vigor. Silkies should not be as tiny as Calls and East Indies. Select birds that resemble wild Mallards in conformation but with medium to medium-long bills and oval heads.

The degree of silkiness varies and is a characteristic that can be improved through selective breeding. Keep in mind that the more extreme the silkiness, the less normal the wing feathers are. The less normal the flight feathers, the greater the likelihood that they will not fold smoothly against the body when the wings are closed. (Silkie chickens share this phenomenon, but they naturally have shorter wing feathers so it is not as obvious.)

The *silky* feather allele Si^s is autosomal, recessive. To have silky plumage, a duck must be homozygous for *silky* (Si^sSi^s). Individuals that are heterozygous at this locus (Si^+Si^s) are normal feathered, but they will produce

A Black Silkie young drake, bred on our farm, with typical breed conformation.

A Gray Silkie young drake, bred on our farm, displaying the unique feathering of this rare breed.

some silky-feathered offspring if paired with either a homozygous or heterozygous mate.

We have found that when birds that are heterozygous for Si^s are used on one side of a mating every second or third generation, better fertility and hatchability are obtained. This is true for normal-feathered heterozygous (Si^+Si^s) drakes mated with homozygous (Si^sSi^s) silky-feathered females and vice versa. The expected ratio of the offspring from this mating is 1:1 silky to normal feathered.

Selecting and preparing show birds. If you are showing Silkies, select birds that are moderately small, have typical conformation, and above all else possess distinctive silky plumage that is sparkling clean. Silkies are naturally tidy, and if you provide them with a balanced diet, a clean pen with sufficient room, fresh bathing water, and plenty of shade, they will basically prepare themselves for shows. To help maximize feather quality and condition, during the molt and until the end of the show season you can use the ration recommended for East Indies.

When sending in the show entry form, you may want to emphasize that these "are ducks." At one show our niece entered her Silkies; the show secretary could not comprehend that these could really be Silkie Ducks, and so for the duration of the show they were penned in the chicken section of the exhibition hall.

At this writing Silkie Ducks are not yet recognized by the American Poultry Association or the American Bantam Association. They can be shown at most poultry shows, but they are not eligible for awards in competition with officially recognized breeds.

Comments. Here in western Oregon, winters are notorious for being long, wet, and cool, but our Silkies have thrived being outside during the day year-round. They have free access to a dry, predator-proof shelter, where they are shut in every night. When temperatures fall into the teens, they do choose to spend more time inside on the well-bedded floor.

The majority of females have strong maternal instincts. As with most skills, they often get better with practice, but we have had first-timers do a fine job of incubating and hatching eggs. If the incubation environment is correct for them, Silkie eggs can be hatched in incubators. To get their eggs to dehydrate sufficiently, we use a wet bulb reading that is 4° to 6°F (2.2 to 3.3°C) lower than what we use for most other duck breeds. Eggshell color can be white, cream, blue, or gray. After breeding and observing Silkies for more than a decade, we have found them to be hardy, active foragers that are easily tamed and adaptable to a surprisingly wide range of environments.

Miniature Silver Appleyard

British waterfowl breeder Reginald Appleyard developed both the large (1930s) and bantam (1940s) versions of Silver Appleyard ducks at his Priory Waterfowl Farm near Bury, St. Edmund, in West Suffolk County, England. According to the *British Waterfowl Standards* (1999), Mr. Appleyard's foundation stock for the miniatures was obtained by mating a White Call and a small Khaki Campbell. It would have taken a number of generations and careful selection to fix the color and type that this handsome breed displays today.

Despite the fact that both the large and bantam versions were given the name Silver Appleyard, the plumage colors and patterns of the two differ both genetically and visually. The genotype of the bantam is *dusky* and *harlequin* (Pat^dPat^d, Li^hLi^h), while the genotype of the large Appleyard is *restricted* and *light* (Pat^RPat^R, $Li^{li}Li^{li}$). As far as I know, all the Miniature Appleyards being raised in North America are of the original genotype developed by Mr. Appleyard.

A flock of young Miniature Silver Appleyards, bred on our farm, consisting of two drakes and three ducks.

In Great Britain it is a different story. For years Tom Bartlett was a leading breeder and promoter of the large Appleyards. In the 1980s Mr. Bartlett sent me some work-in-progress photos of his project to create a miniature duck that would be an exact color replica of the large Appleyard. Following his success, the British Waterfowl Association changed their standard and christened this new version "Silver Appleyard Miniature" and renamed Appleyard's original creation "Silver Bantam."

Description. In conformation the Mini Silver Appleyard is an intermediate between the racy, wild Mallard and the compact Call. The bill is of medium length (1½ to 2 inches [3.75 to 5 cm] long) and medium width. The head is oval, of medium size, and moderately streamlined, without the high forehead and puffy cheeks of the Call duck. The medium-length legs are attached near the center of the body, allowing for nearly horizontal body carriage when the bird is relaxed. The Mini Appleyard is smaller than most wild Mallards, at 30 to 38 ounces (0.9 to 1 kg), and quieter than Calls.

Varieties. As with all varieties that have the *harlequin* gene, the Mini Appleyard produces offspring with more or less pigmentation. Although most fall within the range of the Silver category, there are some that are so dark that the males have little or no neck ring and the sides of their bodies are nearly solid claret — I call these Chestnut Appleyards. (See chapter 10 for color descriptions and information.)

Selecting breeders. Choose active, bright-eyed birds that have good breed type and moderately small size. Mini Appleyards should not be as tiny as Calls or East Indies. The desired plumage color is the same as for Welsh Harlequins. The bills of the females normally are dark green or gray, shaded with varying amounts of orange (rather than the solid greenish black of the Harlequin).

Selecting and preparing show birds. At this writing the Mini Appleyard is not in the APA or ABA standards. They can be shown at most poultry shows, but they cannot compete against standardized breeds. In size and conformation good Mini Appleyards look like they could be a wild duck. As with all the bantam breeds, they are tidy, and if you provide them with a balanced diet, a clean pen with sufficient room, fresh bathing water, and plenty of shade, they will prepare themselves for shows. During the molt and for 6 to 8 weeks prior to a show, use the ration recommended for Calls and Mallards to encourage excellent condition.

Comments. The combination of wonderful aesthetics, engaging personalities, mothering ability, hardiness, and enthusiastic foraging makes the Miniature Appleyard a very useful, ornamental, and enjoyable breed.

5

Lightweight Breeds

DUCKS IN THE LIGHTWEIGHT CLASS weigh between 3½ and 5½ pounds (1.5 and 2.5 kg). They are good to outstanding egg layers. Although some of these breeds are classified as nonsetters, individual birds that will successfully incubate and hatch eggs can be found within all of these breeds. If managed properly, each yields gourmet meat in a portion suitable for two to three people.

The Lightweight breeds, along with some of the Bantams, are the most active foragers among ducks, covering more ground as they eagerly search for and consume slugs, snails, insects, and other edibles. Keepers of large livestock find that ducks can be effective agents for controlling liver fluke infestations. While all ducks enjoy swimming, these breeds are more at home on land than most. Under most circumstances they are not capable of sustained flight. However, they can propel themselves over 2- to 3-foot (0.6 to 0.9 m)-high barriers if startled or hungry. As a group they tend to be high-strung, although there is wide variation from strain to strain and even bird to bird. If dealt with calmly, they normally become friendly.

Drakes in this class tend to have a high libido. In single-male matings a drake will typically fertilize anywhere from four to eight females. In flock matings a ratio of five to seven ducks per drake normally gives good results. To avoid injury to the females, it is important not to have an excessive number of males.

Laying ability and egg size are characteristics that are strongly influenced by the father. Therefore, in order to maintain high yields of large eggs, it is prudent to carefully choose breeding drakes from high-producing families.

If efficient egg production is a principal goal, one of the Lightweight breeds should be given top priority.

Bali

The Bali is one of the oldest domestic ducks. The early histories of the Bali and Runner most likely parallel each other, since the former appears to be essentially a crested version of the latter. Duck herding has been an integral part of traditional rural life on Bali and other Indonesian islands and in various parts of the Southeast Asia mainland for thousands of years. In duck flocks kept in these areas, it was common for crested birds to be among their numbers. In North America and Europe, the crested birds became known as the Bali, Balinese Crested, or Crested Runner; those with plain heads were simply called Runners. The Bali is not recognized by the American Poultry Association (APA) and is only sporadically available.

Description. The main differences between the Bali and Runner are that the former has a chunkier body, is a bit wider across the shoulders, has a coarser head and bill, and has a crest on its head. Standing at attention, typical body carriage is 60 to 75 degrees above horizontal. All other characteristics are similar to those of Runners.

Due to the genetic makeup of crested ducks, when mated together one out of three offspring on average will be plain-headed and two will have crests. (For additional information on breeding ducks with crests, see section on Crested ducks, page 71.) Another characteristic is that ducklings out of the same parents typically have crests that vary widely in size and shape. Crested ducks are prone to produce some offspring with arched necks, deformed backs, or neurological problems.

A White Bali drake displaying typical breed characteristics of a rounder head, thicker neck, and chunkier body than its close relative, the Runner.

Varieties. Balis have been bred in many of the same varieties as Runners. Traditionally, the most common colors were various shades of brown, white, and a modified Mallard pattern. (See chapter 10 for color descriptions and information.)

Selecting breeders. First and foremost, select birds that have strong legs and an abundance of vitality. Do not breed from birds with balance problems, deformed backs, or severely kinked necks. Be aware, though, that because arched necks are so prevalent in crested ducks, breeding from such specimens is sometimes unavoidable.

When possible, choose birds with smooth, well-rounded crests that are attached as high on the head as possible. For breeding birds, the size of the crest is not critical. In fact, those with smaller crests are usually more productive and will normally produce some offspring with significantly larger headgear.

The way to produce the largest number of offspring with the fewest deformities is to mate crested to noncrested birds. This mating, which was the usual procedure in the homeland of the Bali, typically produces half crested and half noncrested ducklings. Avoid breeding from birds with short or dished (concave) bills, large round heads, and thick, heavyset bodies with low body carriage.

Selecting and preparing show birds. Look for graceful, moderately slender, upstanding birds with good height, straight backs, and medium-size crests that are as balanced and round as possible. When transporting them to shows, use containers that are taller than the birds to avoid temporary kinked necks. During the molt and for 6 to 8 weeks prior to a show, a good feeding regime for encouraging excellent condition without excessive weight is to supply oats (whole or pelleted) and granite grit free choice. In addition, feed a mix of seven parts (by volume) game-bird flight conditioner and one part cat kibbles morning and night in a quantity that they will clean up in 10 to 15 minutes.

Comments. In my opinion, the Bali is the most unusual of all breeds. With their dignified yet comical appearance, they can bring a smile to the face of even the most serious-minded person. They are generally good layers of white, green, or blue eggs. Locating specimens to get started with is a worthwhile task, as Bali is a fascinating breed to raise.

Campbell

Mrs. Adele Campbell of Uley, Gloucestershire, England, originated this breed in the late 1800s. In a book titled *Ducks — Show and Utility*, C. A. House included the following information on the early history of the breed:

> Writing to me on the origin of the breed, Mrs. Campbell says: "The real beginning of the Khaki Campbell was a great appetite on the part of my husband and son for roast duckling! To make ducklings, one must have duck eggs, and I had just one duck, a Fawn & White Indian Runner,* which laid 195 eggs in 197 days. She was the only duck in the yard, a rather poor specimen in appearance, and no pedigree. However, I thought some good layers might be expected from her, but I wanted a little more size and mated her to a Rouen drake. . . . The original Campbells were practically this cross except that one season a Mallard drake was used. . . ."

From this humble foundation one of the world's most prolific egg-laying birds was developed.

Introduced to the public in 1898, the original Campbells resembled poorly colored Mallards. Drakes had dark green heads, ill-defined neck rings, grayish bodies, and pale claret breasts. Ducks were grayish-brown, penciled with dark brown, and often had a patch of white on the neck.

In an attempt to make a buff-colored duck, Mrs. Campbell mated her original Campbells back to Penciled Runners. Rather than the sought-after buff-plumaged birds, she ended up with tannish-brown ducks that reminded her of the khaki-colored uniforms worn by the British soldiers fighting in the South African War of 1899. The Khaki variety, which was introduced in 1901, grew in popularity and the original color variety gradually disappeared.

The ability of the Campbell to produce prodigious quantities of eggs was evident early in its history. Prior to 1920 reports had surfaced in British poultry magazines and books of small flocks of Campbells that averaged close to 300 eggs per duck per year. The record for an individual duck of that era was a mind-boggling 360 eggs laid in 365 days.

---

* The variety called Fawn & White in Great Britain is known as Penciled in North America. The American Fawn & White is a different color and is not recognized in England.

The most successful large-scale breeder in the history of Campbells was Mr. Aalt Jansen of Ermelo, Holland. Mr. Jansen was a fisherman who began raising Khaki Campbells in 1921. By 1950 the famous Jansen's Duck Breeding Farm had approximately 50,000 birds that averaged 335 to 340 eggs per duck at seventeen months of age. One of their amazing Campbells began laying at 108 days of age and laid an egg a day for 405 consecutive days. The Jansens' success was based on their pedigreed breeding program, carefully formulated feeds, and outstanding farm management.

The earliest record that I have found of Campbells in the United States is the importation from England of nine birds in 1929 by Perry Fish of Syracuse, New York. The Khaki Campbell was admitted to the *American Standard of Perfection* in 1941.

After languishing in relative obscurity in North America, Campbells began to be raised in greater numbers during the late 1970s. The primary causes for this upswing were threefold: the back-to-the-land movement that started in the 1960s, the surge of Asian immigrants (many of whom relished duck eggs) at the end of the Vietnam War in 1975, and the importation of high-producing Campbells from the Kortlang Duck Farm of England in 1977.

Today most Campbells continue to be raised for their wonderful practical qualities, although fine exhibition specimens can be seen at many of the larger poultry shows.

Description. In conformation this is a breed of moderation. The late Henry K. Miller of Pennsylvania, a longtime Campbell breeder and respected poultry judge, told me on a visit to his farm, "Campbells are built the way a duck should be built." They are active, moderately streamlined birds. The head, bill, neck, and body are all modestly long, with a sprightly body carriage of 20 to 40 degrees above horizontal. When the Khaki is in prime feather condition, the warm earth tones of its feathers make it a lovely duck to look at, especially when seen on a carpet of green grass.

A quirk in the genetic makeup of Khaki and Dark Campbells results in some of the offspring hatching with a bit of white on the front of the neck and/or under the bill, even if only solid-colored birds have been carefully selected for breeding stock for many generations.

Varieties. Khaki is the main variety. Whites were derived as a "sport" from the Khaki and are raised in small numbers both in England and North America. The Dark Campbell, which in North America is rare, was originated by Mr. H. R. S. Humphrey of Devon, England, in order to make possible the production of sex-linked Campbell ducklings. Dark Campbells have the same

A young pair of Khaki Campbells bred on our farm. Females from this production-bred strain typically produce 320 or more eggs during their first year of lay.

plumage pattern as Khakis, but they are considerably darker due to their lack of the *sex-linked brown* allele. (When a Khaki drake is mated to a Dark female, the resulting offspring are all Dark males and all Khaki females. The reciprocal cross is not sex-linked.) Pied Campbells, which have the same pattern and color as Penciled Runners, were bred out of pure Campbell stock in Oregon in the early 1980s. (See chapter 10 for color descriptions and information.)

Selecting breeders. To maintain high production in laying ducks, it is critical to choose breeding birds that are robust and bright-eyed and have strong legs. Drakes should be sons of prolific dams that produce strong-shelled eggs of the desired size.

In the Northern Hemisphere, especially above the 30-degree parallel, the normal time for birds of most species to lay is during the spring, when both daylight hours and temperatures are increasing. As Leslie Bonnet stated succinctly in his book *Practical Duck-keeping* (1960), "In the spring even sparrows lay." Therefore, it is prudent to select breeder ducks that lay out of season, especially during the months when daylight hours are the shortest and the weather the most inclement.

Longevity and lifetime egg production are valuable characteristics to select for. In our breeding program we prefer to use ducks that have sustained high egg yields for 4 or more years. It is not unusual for these females to have lifetime production records of more than 1,200 eggs. Many of our best annual and lifetime layers have been produced by two-way or four-way line crosses (a line is a carefully inbred family of birds). When two or more inbred lines of ducks are crossed, their offspring typically have improved production.

Some sources have stated that ducks with certain physical characteristics (such as long bills with level toplines, eyes that are set high in the head, and long bodies) are the best layers. After raising Campbells for more than four decades, I have not found this to be true. Some of our best layers (including a duck that laid 357 eggs her first year of production and went on to produce many excellent sires and dams) possess average-length bills with concave toplines, eyes that appeared to be set relatively low in the head, and medium-length bodies. In my experience the laying potential of a duck cannot be reliably judged by external characteristics.

Selecting and preparing show birds. Most judges prefer sleek birds with moderately long bodies, nearly straight necks of medium length, refined heads, and medium-length bills with a nearly level topline. Body carriage should be intermediate between the upright Runner and the horizontal Call (about 30 degrees). Khakis should have no white feathers visible when they are in the show cage and must not have clear yellow bills. Some judges also check along the jawline and under the bill for white feathers.

Because khaki plumage is susceptible to fading, successful exhibitors often keep their birds in the shade when the sun's rays are the strongest. During the molt and 6 to 8 weeks prior to a show, you can use the same feeding recommendations as outlined for Bali ducks to encourage excellent condition without excessive weight.

Comments. If they consume an adequate diet, are kept calm, are provided sufficient space, and run in flocks consisting of no more than an ideal of 50 to a maximum of 200 birds, Campbells have proven to be amazingly adaptable. They have performed admirably in environments ranging from arid deserts with temperatures of 100°F (38°C) to humid tropical rainforests with more than 200 inches of annual precipitation to cold northern regions where temperatures can remain below 0°F (–18°C) for weeks at a time.

The egg-bred Campbell, along with its derivative the Welsh Harlequin, is the best layer of all recognized duck breeds and in many environments is the most proficient egg producer of all avian species. Campbells lay eggs with

> ### A WORD OF CAUTION
>
> If you want good egg yields, make certain you acquire authentic Campbells that have been selected for high egg production. The laying ability of birds in some flocks has been allowed to deteriorate, and, much too often, crossbreds are being sold as Campbells. Any alleged Campbell that has facial stripes or weighs more than 6 pounds (2.75 kg) is a crossbred.

superb texture and flavor and usually with pearly white shells. However, an occasional bird of authentic bloodlines will lay eggs with blue or green shells. I consider Campbells, Welsh Harlequins, and hybrids sired by either of these to be the best of all birds for the efficient production of eating eggs. Campbell drakes from a strain selected for high egg production make outstanding sires for the production of hybrid laying ducks.

Dutch Hook Bill

Holland is famous for its waterways, so it is not surprising that duck culture has a long and colorful history there. During visits to our farm, poultry specialist Hans L. Schippers of Holland told me about the rare, old Dutch Hook Bill ducks that had a close encounter with extinction in the mid-1900s. According to his account and references he provided, the Hook Bill has been traced back as early as 1676. At one time large numbers of Hook Bills were raised for their eggs and meat in the Netherlands, where they were allowed to forage in and along the canals. Tradition holds that the distinctive bill shape and white bib were selected for to make it possible for hunters to distinguish them easily from wild ducks. Mr. Shippers says the typical weight of Hook Bills in Holland is 5 to 6½ pounds (2.25 to 3 kg), which is heavier than the typical weights of those raised in Great Britain (4 to 5½ pounds [1.75 to 2.5 kg]). After raising and observing Hook Bills for the last decade, I consider them to be one of — if not the best — forager of any domestic duck breed.

Description. The first characteristic that catches your eye is the unique shape of the head and bill. In most breeds the topline of the bill is slightly concave and the forehead rises noticeably at the junction of head and bill. However, the Hook Bill head is streamlined and flows smoothly into a long bill that, in extreme cases, is strongly curved downward. Upon further observation it can be seen that the body shape tends to be tubular and they have longer legs (in proportion to their size) and tighter plumage than most breeds. When they are relaxed, the topline of their body is approximately 20 degrees above horizontal, but when they are excited, it can go to 45 degrees or more.

The bills of most colored Hook Bill drakes are a striking blue-gray. Those of females are either pinkish orange, usually with a dark/gray chevron-shaped saddle mark, or a solid grayish-black. Most self-Whites of both genders have pink bills, with females typically developing more or less black spotting as they near egg production. Because the genetic diversity of Hook Bills is limited, we have used minimal selection pressure for bill color.

Varieties. The traditional colors include Dusky, White-Bibbed Dusky, and solid White. (See chapter 10 for color descriptions and information.)

Selecting breeders. The first year we raised Hook Bills, we observed that there was noticeable variation in bill shape even among full siblings. We identified three groups based on bill shape: (1) extreme downward curve, (2) moderate downward curve, and (3) straight. The following breeding season we tried different kinds of matings, including (A) moderate × moderate, (B) straight × extreme, (C) moderate × extreme, and (D) extreme × extreme. The results were instructive: The A matings produced ducklings with all three bill-shape categories, and the eggs hatched well. The B matings produced ducklings with all three bill-shape categories but predominantly with moderate curvature, and the eggs hatched well. The C matings produced ducklings primarily with moderate and extreme curvature, and the eggs hatched fair. A majority of the ducklings from the D matings had extreme bill curvature, and the eggs hatched poorly. Because our main interest is in perpetuating the Hook Bill and maintaining its useful qualities, we primarily use the A- and B-type matings in our breeding program.

Our six-month-old White-Bibbed Dusky Dutch Hook Bill drake, bred from stock from Great Britain, has a well-defined bib and a moderately downward curved bill.

Our studies of the inheritance of the hook bill trait show it to be dominant to wild type and with a wide range of expressivity. Major genetic control may occur at multiple loci. Full-term embryos that fail to hatch frequently have bills with exaggerated curvature and/or are abnormally short.

When our first Hook Bills arrived from quarantine, their calm disposition was a surprise. Because personality traits are somewhat inheritable, we choose breeding birds that possess the calmness for which the breed is known.

Selecting and preparing show birds. The head and bill shape is such a distinctive feature of this breed that it should be taken into account when selecting birds to show. However, both breeders and judges would be advised to not go overboard on the emphasis of these features to the detriment of the breed. In the Bibbed variety choose birds with well-defined bibs (exact size of the bib is not critical). It is within show rules to remove stray white or colored body feathers to make the markings more distinct. Hook Bills kept in a clean environment and provided bathing water will basically prepare themselves for showing.

Comments. The unique shape of the bill becomes more obvious after ducklings are several weeks old. We have found the Hook Bills to be good layers — most females lay blue eggs, but a few lay white eggs. With their tremendous foraging ability, they are extremely useful in controlling many garden and pasture pests. Some people have been using them very successfully in two-, three-, and four-way hybrid crosses to produce "designer" ducks for specific uses. These crosses (often with Runners, Campbells, and/or Harlequins) have hybrid vigor and are agile, excellent layers, and exceedingly good foragers. Some are designed specifically for training herding dogs. While thought to be extinct at one point, Hook Bills survived — thanks to a few dedicated breeders in post–World War II Europe.

Magpie

Oliver Drake and M. C. Gower-Williams of Wales are credited with developing this charming duck. Magpies are not mentioned in publications prior to the early 1900s, indicating that either they were not widely known or did not yet exist. From its conformation, and the genetic composition of its plumage pattern, the Magpie likely has Runner in its ancestry.

One interesting possibility is that the Magpies descended from the Huttegem, an old Belgian duck that was raised in great numbers during the 1800s in the duck growing district near Ondenaarde, 30 miles west of Brussels. The Huttegem is thought to have been developed from the Termonde and Runner breeds. Old pictures of the Huttegem clearly show birds that genetically carry *dominant bib* and *runner pattern* genes, which means they would have produced some offspring with Magpie markings. Furthermore, in Edward Brown's *Poultry Breeding and Production* (1929), his shape description of the Huttegem's head, bill, body, and station is an amazingly accurate portrayal of the Magpie.

According to Darrel Sheraw's book *Successful Duck and Goose Raising* (1975), Magpies were imported from Great Britain in 1963 by the late Isaac R. Hunter of Michigan. Following this importation, a handful of breeders — Mr. Sherraw, Jim Cleaver, and Curtis Oakes, all of Pennsylvania, and the Urch family of Minnesota — were instrumental in keeping the Magpie alive in North America. In 1977 the Magpie received a boost when it was admitted to the *American Standard of Perfection*. Starting in 1984, Magpies became more readily available and since then have been gradually increasing in numbers.

A factor that hindered the popularity of Magpies among hobbyists was the virtual impossibility of meeting the specifications for bill and leg color as written in both the British and American standards. In 1998, upon my recommendation, the APA revised its standard specifications to bring them in line with the genetic potential of the Magpie, which should encourage more specialty breeders to take up this worthy breed.

A Black & White Magpie old drake with exceptional plumage markings. Bred on our farm, he was the winner of Reserve Champion Light Duck at the 1993 APA National.

Description. The name of this breed refers to its distinctive markings. In the ideal specimen the plumage is predominantly white, offset by two colored areas — the back (from the shoulders to the tail) and the crown of the head. Breeding well-marked Magpies is a constant but highly rewarding challenge. Even out of the best matings, only some of the offspring will have good exhibition markings. However, a great advantage of the Magpie is that the best-marked ducklings can be identified at hatching. The well-marked birds can be raised for exhibition while the others can be raised for breeding, pets, egg production, or meat. An interesting characteristic of Magpies is that as they age the mantle and cap, especially in ducks, will typically become mottled with white and may eventually turn totally white.

In conformation and station, the Magpie resembles the Campbell and Harlequin. The American weight specification for old drakes is 5 pounds (2.25 kg), whereas in Great Britain it is 5½ to 7 pounds (2.5 to 3 kg). American-bred Magpies vary in size, and anyone who prefers a larger bird could select a strain that would weigh 6 to 7 pounds (2.75 to 3 kg). Because of their white underbody, they dress off almost as cleanly as solid white ducks.

Varieties. Black and Blue are APA recognized colors. Silvers are a natural product of Blues; Chocolates are extremely rare. Most matings of Magpies will produce some pure white offspring, traditionally called Stanbridge Whites. (See chapter 10 for color descriptions and information.)

Selecting breeders. Birds used for breeding should be robust, strong-legged, free of physical deformities, good layers, and of correct body size and type. A mistake most beginners make with Magpies is discarding many of the birds that make the best breeders. Magpies are a breed in which mating together the best show birds often does not produce the best-colored offspring. (See chapter 10 for details on mating Magpies for color.)

Selecting and preparing show birds. Good show birds have caps that cover at least half of the crown of the head and a clearly defined back mantle. Because the Magpie pattern is a challenge to perfect, the main emphasis by both exhibitors and judges should be clearly defined markings that are symmetrical rather than the exact size or shape of the markings. The bill is yellow or orange in young birds, gradually turning green with age. A long, sleek body with a nicely rounded chest, good width between the legs, and a body carriage of approximately 25 degrees above horizontal are desired. A medium-long neck with a moderately racy head and bill finish off the ideal show bird.

As with all marked varieties of show birds, most Magpies benefit from the removal of stray off-colored small feathers in the head, neck, and body plumage a day or two before a show. However, the removal of large tail or wing feathers is not permitted. During the molt and for 6 to 8 weeks prior to a show, you can use the same feeding recommendations as outlined for Bali to encourage excellent condition without excessive weight.

Comments. In 1987 we sent 500 ducklings to Guatemala. These ducks were distributed to Kekchi villagers to help alleviate the high incidence of malnutrition among their children. The eggs and meat provided essential nutrients that were lacking in the local diet. (The ducks performed admirably in this wet, humid region that was inhospitable to chickens.). The Magpies were the villagers' favorites because of their all-around performance and eye-catching markings (see photo on page 58).

Magpies are a true triple-duty duck. Along with being highly decorative, they are wonderful layers of large (white- or bluish-shelled) eggs and produce gourmet meat. Everyone who raises them can have the satisfaction of knowing they are helping perpetuate one of the rarest standard breeds of ducks.

Runner

Modern Runners are descendants of the traditional herding ducks raised along the coastal regions of the Indo-Chinese Peninsula and the islands of Southeast Asia. Hieroglyphics have been found in ancient Javan temples that suggest Runner-type ducks have existed for more than 2,000 years.

Raising and herding ducks has been a tradition in parts of Asia for many centuries. One form of duck herding consists of herders taking their ducks out to rice paddies and fields during the day where the birds glean shattered grain, weed seeds, snails, insects, larvae, small reptiles, and such. The herder may carry a long pole, often adorned with a flag or some feathers, which the ducks recognize. When their destination is reached, the pole is stuck in the ground, and the ducks learn to stay within sight of the guidepost.

At day's end the herders either lead or follow their charges home, where the ducks are typically enclosed in a bamboo or clay encampment. Because this type of duck normally lays in the early-morning hours, their eggs are easily gathered prior to the foray into the countryside. Over their long history of being herded, a duck evolved that was a deft forager and could travel long distances at a quick pace.

Tradition has it that Runners were introduced into the United Kingdom from Malaya by a ship's captain around 1850. These birds landed at White-haven on the northwest coast of England and were bred by friends of the captain in Cumberland County. Many were blue with white markings or brown with white, the latter possibly forerunners to the Penciled variety.

Before long they had spread to farmers in the adjoining county of Dumfries in Scotland. According to the *British Waterfowl Standards* (1982), fawn-colored Runners were shown at the Dumfries Show in 1876, Penciled were shown in 1896, and the Black and Chocolate varieties were included in the written *Standard* of 1926.

In North America, after considerable controversy over what color "true" Runners were, the Fawn & White variety was admitted to the *American Standard of Perfection* in 1898. Finally, after cooler heads prevailed and it was recognized that there could be more than one color of Runners, both the pure White and the Penciled varieties were standardized in 1914. The Black, Buff, Chocolate, Cumberland Blue, and Gray varieties were admitted to the *American Standard* in 1977.

Due to their unique appearance and high egg production, a Runner boom swept across Great Britain, France, Germany, and North America. Books and hundreds of articles were written on Runners, and their laying prowess was widely advertised and often exaggerated. Demand was so high that crossbreeding was common, and any webfoot remotely resembling a Runner was peddled as the real article. Authentic Runners were saved by a few dedicated breeders and by adding new blood with direct imports from Asia as late as 1924. The boom inevitably dissipated, but Runners remain popular because of their practical qualities and unique appearance. At many shows Runners constitute one of the largest classes of ducks and win their fair share of top awards.

Description. Of all the domestic ducks, the Runner departs the furthest from its Mallard ancestors in type and station. Runners are a splendid example of what selective breeding can accomplish. Proper Runner type is often graphically described as a wine bottle with a head and legs. When relaxed and strolling about their home range, Runners typically have a posture that is 45 to 75 degrees above horizontal. When startled or standing at attention, good specimens will strike a nearly perpendicular pose. Perched on top of a long, slender, and straight neck is a slim, wedge-shaped head and bill. The eyes are set higher in the head than in any other breed. When excited, good Runners will hold their tails straight down in line with their back. When relaxed, even the best show birds normally hold their tails somewhat elevated. In contrast to the waddle of most ducks, a classic trait of the Runner is a smooth running gait.

Varieties. Historically, Runners have been bred in more color varieties than any other breed. Standard varieties in North America include Fawn & White, Penciled, White, Black, Buff, Chocolate, Cumberland Blue, and Gray. Nonstandard varieties include Faery Fawn (one of the original varieties from Asia, it resembles muted Mallard colors), Blue Faery Fawn, Golden, Saxony, Blue Fawn, Pastel, Trout, Dusky, Khaki, Cinnamon, Silver, Lavender, Lilac, Blue-Brown Penciled, Blue Fawn Penciled, Emery Penciled, Porcelain Penciled, and Splashed. New varieties continue to be developed. (See chapter 10 for color descriptions and information.)

Selecting breeders. In all breeds there is a tendency for some offspring in each generation to regress slightly toward the original type of their wild ancestors. Therefore, to perpetuate the classic characteristics of the Runner, breeding stock must be carefully chosen.

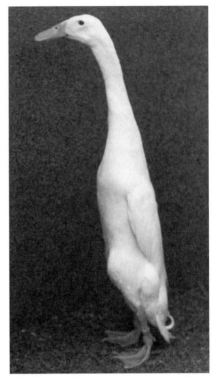

A White Runner old drake bred on our farm with the classic wine-bottle shape. He won Best Bird in Show at the Washington Feather Fanciers 1994 Autumn Classic.

Faults to avoid include low body station; short, stocky bodies with prominent shoulders and chests; short necks; round heads with prominent foreheads; short bills and/or concave topline; and tails that are constantly cocked upward, even when the bird is excited (many Runners will hold their tails down better in the unfamiliar surroundings of a show than they do at home). Birds that always hold their tails straight down are not necessarily desirable either, because this can be a sign of leg weakness, and they tend to soil themselves. Bills want to regress to the original concave shape, so it is useful to select some breeders that have powerful bills with a slightly convex topline.

Look for strong legs and a smooth running gait. Keep in mind that until their muscle tone is well developed, juvenile Runners can have shaky legs or sit down frequently when they are exercising more than they are accustomed to. However, a mature duck that does not move about freely under normal circumstances should not be used for breeding purposes. Remember: Any duck that is terrified or runs harder than its legs are conditioned for will likely temporarily lose its ability to walk normally from fatigue or muscle strain.

Selecting and preparing show birds. Many of the better shows have Runners judged in a ring rather than cages. This method has the advantage of allowing the running gait of the contestants to be evaluated, eliminating any possibility that a bird with weak legs will win. Some Runners pose the best in cages, but others show off to their best advantage in the ring. Keep in mind that some Runners that look quite average at home show off like champs when put in the environment of the exhibition hall. Until a bird has been shown several times, it can be difficult to evaluate its true show worth.

An outstanding exhibition Runner is smooth, slender, and nearly vertical, with an imaginary straight line running from the back of the head through the neck and body to the tip of the tail. Everything else being equal, the taller the bird and the longer and straighter its bill, the better.

Good Runners are genetically slender and do not have to be fed dangerously small quantities of feed to maintain slim profiles. For mature Runners, during the molt and for 6 to 8 weeks prior to a show, follow the feeding regime recommended for Bali. To encourage adequate exercise, our grassy pens are long, with waterers and feeders located at opposite ends. Runners fed unlimited quantities of a fattening diet, such as corn, meat-bird grower, broiler grower, or cat food often become obese and do not show well.

To avoid kinked necks, pen Runners in cages that allow them to stand upright without bumping their heads on the top. The minimum height of show cages used for Runners is 27 inches (69 cm). The carrying boxes we transport our Runners in are 32 inches (80 cm) tall to allow room for an abundant layer of bedding in the bottom.

Runners are the tightest feathered of all ducks, so it is easy for their wing feathers to become disheveled during transportation or as they are placed into show cages. Prior to judging, make sure their flight feathers are folded properly.

Comments. The graceful Runner is one of the most entertaining and useful members of the duck clan. When she sees a group of our Black Runners marching in a line, a friend of ours likes to declare, "Look, the miniature monks are going to town."

From my observations Runners (along with Dutch Hook Bills and some bantam breeds) are the most active foragers and will cover the largest territory in their hunt for snails, slugs, insects, and other edibles. Their active disposition is evident right from the start. When taking newly hatched Runner ducklings from the incubator, one must move slowly and talk to them quietly to keep them from jumping overboard in their enthusiasm to explore their expanding world.

Runners make elegant show birds, entertaining pets, and wonderful pest controllers, and they are fine layers. Herding-dog trainers often start their young trainees on Runners. Without a doubt, Runners deserve their wide popularity.

Welsh Harlequin

Leslie Bonnet was a well-known commercial duck breeder who lived near Criccieth, Wales. In his book *Practical Duck-keeping* (1960), he wrote the following about the breed he developed:

> The Welsh Harlequin originated from two sports of pure Khaki Campbell stock in 1949. Its supporters claim that the breed is a better egg producer than the Khaki Campbell. If this is so, it would be due to the docile and placid nature of the breed, which reduces chances of interruption of egg-laying through shocks or scares. . . . A flock averages over 300 eggs per year.

John Fugate, of Tennessee, imported hatching eggs from Bonnet in 1968 with the hope of introducing this unique breed to North America. By 1981 the descendants of these original imports had dwindled to two tiny flocks. Concerned that his effort was failing, John contacted Millie and me, asking if we would help him establish Harlequins in North America. We agreed and set up pedigreed matings to help evaluate and maximize the gene pool of the Harlequins he sent us. To increase genetic diversity John imported adults in late 1981. By 1984 we had sufficient matings to be able to ship ducklings to interested parties throughout the Americas.

Early in our work with the Harlequins, I noticed that there were two distinct colors and that the body conformation tended to be significantly shorter than portrayed in the earliest descriptions and photos of the breed. These changes indicate that Bonnet introduced non–Khaki Campbell blood into his breeding program at some point prior to 1968. John and I decided that we would select for the longer body conformation of the original Harlequins but would maintain both color varieties. I dubbed these Gold and Silver. During the 1990s several British Waterfowl breeders and judges who visited us com-

An old pair of Silver Welsh Harlequins bred on our farm with typical type and color markings. Although difficult to see in a black-and-white photo, the female's plumage is shaded with soft fawn.

mented that the conformation, size, plumage pattern, and bill and leg color of our Harlequins were more authentic than many of those found at that time in their homeland.

Harlequins are primarily raised for their wonderful practical attributes. However, they are also stunningly beautiful ducks, are highly decorative, and make dazzling show birds. The qualifying meet for admission into the American Poultry Association (APA) *Standard of Perfection* was held in October 2001, at the Western Waterfowl Exposition in Albany, Oregon, with 74 Harlequins shown.

Description. In conformation the Harlequin closely resembles the Campbell. They are moderately streamlined throughout, with relatively long bodies, medium-width backs, nicely rounded chests, smooth underlines, and sufficient width between their legs to allow good capacity in the moderately full abdomens. Their necks are of medium length and held nearly vertical. Heads are trim and oval-shaped with medium-long bills that are straight or only slightly concave along their topline. The Harlequin body carriage is a sprightly 20 to 35 degrees above horizontal, augmenting their active foraging habits. The resplendent colors and pattern of the Harlequin make it difficult to comprehend that the genetic differences between their plumage and that of the Khaki and Dark Campbell are the result of a single pair of genes.

Varieties. The original color was the Gold, which has no black pigment, has soft colors, and is the equivalent of Khaki in Campbells. The Silver variety apparently arose at least 10 years after the origin of the breed and has the same relationship to the Gold variety as Dark has to Khaki in Campbells. Silver Harlequins have more contrast and brilliance in their plumage and are the most common variety in North America today. In Great Britain the Silver variety is not recognized as a Harlequin. (See chapter 10 for more on color.)

Selecting breeders. Birds used for breeders should be robust, strong-legged, free of physical deformities, heavy layers, and of correct body type and color. To help perpetuate the authentic Harlequin, avoid the following characteristics: more than ¾ pound (0.3 kg) above or below typical weights; short, blocky bodies; large, coarse heads; Mallard-like facial stripes; light-colored bills in ducks; and poor producers.

Selecting and preparing show birds. In size, type, and stance, the ideal show Harlequin looks like a slightly longer and larger Campbell. The size difference between these two breeds is approximately a half pound (0.25 kg), which is a 10 percent difference. During the molt and for 6 to 8 weeks prior to a show, you can use the same feeding recommendations as outlined for Bali to encourage excellent condition without excessive weight.

Comments. When we took our first tray of Harlequin ducklings out of the hatcher in 1983, there were several distinct bill colors. Upon vent sexing the hatchlings, we discovered that those with the darker bills were almost all males and those with light-colored bills ending in a dark tip were females. After vent sexing thousands of Harlequin ducklings during the subsequent years, I have found that they can be sexed with at least 75 percent accuracy by day-old bill color. A few days after hatching, the distinction begins to disappear.

Harlequins have proven to be one of the most important additions to the North American duck roster in the past 60 years. They are beautifully colored, highly adaptable, outstanding layers, active foragers, excellent producers of lean meat, and because of their light under-color, they pluck almost as cleanly as white birds when dressed for meat. Even though their egg production is near or equal to bred-to-lay Campbells, a fair number of the females will successfully incubate and hatch eggs. Most Harlequins lay eggs with pearly white shells, although a minority lay colored eggs. Harlequins richly deserve their growing popularity. Harlequin drakes from a strain selected for high egg production make outstanding sires for the production of hybrid laying ducks.

6

Mediumweight Breeds

DUCKS IN THE MEDIUMWEIGHT CLASS typically weigh 6 to 8 pounds (2.75 to 3.5 kg). They are general-purpose ducks, being fair to good layers, moderately fast growers, and producers of high-quality meat that is more flavorful and less fatty than that of most Pekins. When raised as broiler ducklings, they typically are feathered out under the wings and ready for butchering 2 to 3 weeks later than the fastest-maturing strains of Pekins. These breeds are well suited for situations where they can forage for some of their own food and are then butchered once they have their full adult plumage at sixteen to twenty weeks of age.

Mediumweight breeds are good foragers that are capable of eating large banana slugs. For many situations they make excellent yard or pond ducks since they tend to stay close to home, do not fly under normal conditions, and are large enough that they are less likely than Bantam breeds to be preyed upon by winged predators. Typically, they have moderately calm temperaments and make fine pets. They are sometimes the first ducks used to train novice herding dogs.

In single-male matings, a vigorous one- to two-year-old drake will normally successfully fertilize somewhere between three and five ducks. In flock matings four to six ducks per drake is typical. Many ducks of these breeds will get broody and hatch eggs if given a sheltered nesting area that is protected from predators.

Ancona

Most poultry raisers associate Ancona with the breed of chicken by that name. However, during the early part of the twentieth century, a breed of duck was developed in Great Britain and also called Ancona. They were likely developed from the same foundation stock as the Magpie, and the genetic makeup of their plumage pattern supports this hypothesis. Some Anconas look identical to the old Belgian Huttegem duck.

Ancona ducks have been raised in the United States since at least the 1970s and were first exhibited in 1983 in Oregon. Although still not common, their numbers have been increasing steadily since 1984, when they first became available to the public. As of this writing, they have not been officially recognized by the American Poultry Association.

Description. This breed shares many physical characteristics with its apparent close relative, the Magpie. The main difference is the color pattern of their plumage, and Anconas are a little heavier, and their conformation is a bit stockier. The Ancona has a medium-size oval head; medium-length bill that is slightly concave along the topline; average-length and -diameter neck that is only slightly arched forward; and medium-length body carriage of 20 to 30 degrees above horizontal when relaxed.

The broken plumage pattern of Anconas is unique among ducks. Like Pinto horses and Holstein cattle, there is no set design. Any combination of white and color is acceptable as long as there are obvious broken areas on the head, back, sides, and underbody. The neck is normally solid white. At sexual maturity the bill typically is spotted with varying amounts of green, often darkening to solid color with age. The legs and feet are orange with black or brown markings that increase with age.

Varieties. The classic Ancona varieties are Black & White, Blue & White, Chocolate & White, Silver & White, and Lavender & White. Most Anconas will also occasionally produce one or more of the following variations: Fawn & White, Penciled, Pied, and pure White. (See chapter 10 for color descriptions and information.)

Selecting breeders. As with all rare breeds, it is especially important to choose stock birds that are vigorous, are free of physical deformities, and have classic breed traits. Since the Ancona is an excellent laying breed, productivity should be given high priority in breeders. To produce the highest percentage of offspring with the unique broken pattern, select birds with definite colored areas under their eyes and at least a bit of color on their chests. Avoid specimens that are either solid white or primarily colored with a white bib.

Our old trio of Black & White Anconas, showing the body type and haphazardly broken plumage pattern typical of this breed.

Selecting and preparing show birds. Because the most distinctive feature of the Ancona is its broken-plumage pattern, choose show birds that have the boldest patchwork pattern. Avoid birds that resemble Magpies or Swedish.

Comments. The patchwork-marked day-olds are adorable and often are the first ones chosen when people are selecting from a group of assorted ducklings. Because their plumage designs are so diverse, seldom are two Anconas identical, making it easy to distinguish individuals in a flock.

Anconas are the best foragers and the most prolific layers of the Mediumweight breeds. Eggshell color can be white, tinted, blue, green, or spotted. They have proven to be extremely hardy and adapt to many environments.

Cayuga

In the heart of the Finger Lakes region of west central New York lies a slender, 40-mile-long glacial lake named after the Native Cayuga people. By 1863 the big black ducks being raised in the area surrounding the lake were called Cayugas. However, the seed stock for these ducks apparently came from southeastern New York, near its border with Connecticut.

A traditional account of the origin of the Cayuga says that a miller living in Dutchess County, New York, caught a pair of dark-colored ducks on his millpond in 1809. Descendants of this pair were taken to Orange County, New York, where they multiplied. In approximately 1840 John S. Clark introduced them to Cayuga County, where they thrived and gained a reputation as quiet, prolific layers that produced large meat birds.

In 1874 Cayugas were included in the first *American Standard of Perfection*. They were raised in significant numbers as general farm ducks but by the 1890s were largely replaced by Pekins for the market duckling trade in the big cities. Today, Cayugas are raised in modest numbers for pets, decoration, meat, and eggs. The best exhibition strains are of superb quality and frequently win top honors. At large shows Cayugas are often the most numerous of the Mediumweight breeds.

Many writers have suggested that the Cayuga is a descendant of the wild American Black Duck (*Anas rubripes*), which inhabits much of the eastern portion of North America. I have found no hard evidence to substantiate this theory. American Black Ducks are black in name only; their plumage is very dark brown with most feathers edged with lighter brown. The true black plumage of the Cayuga is caused by a black mutation that is fairly common in Mallard derivatives. Furthermore, *Anas rubripes* drakes do not have curled sex feathers in their tails, whereas Mallards and their descendants, including Cayugas, have this diagnostic marker. Until DNA tests are done, it will not be known with certainty whether or not Cayugas have any American Black Duck in their ancestry.

Description. Cayugas have typical duck conformation. Their bodies are moderately long, medium-wide, with good depth carried from the well-rounded chest to the moderately full abdomen. They have necks that are medium in length and diameter and only slightly arched forward. The medium-sized head is oval and the moderately long bill is slightly concave along its topline. Body carriage is approximately 20 degrees above horizontal. The crowning glory of the Cayuga is its resplendent green-black plumage. (See chapter 10 for full discussion of black color.)

Varieties. Black with green iridescence is the traditional color and is the only Cayuga color recognized by the American Poultry Association. Cayugas with solid blue plumage were developed and first shown in Oregon in 1984. (See chapter 10 for color descriptions and information.)

This Black Cayuga old drake that we bred won Grand Champion Duck at the 1991 APA National.

Selecting breeders. Choose birds that are vigorous, strong-legged, and productive. Because many Cayugas are undersized, breeders that have good size, wide backs, and moderately long bodies with thick chests are highly valued. The bill, head, and neck should be in balance with a medium-size duck. Some Cayugas have a row of feathers running down the outside of their legs. There are people who will not breed from these ducks; others will, if they are exceptional specimens otherwise.

Selecting and preparing show birds. Look for ducks that have solidly built bodies and good size. Everything else being equal, the greener the plumage and the blacker the bill, legs, and feet, the better. (Since long exposure to bright sunlight can cause the plumage to develop purple reflections, black show ducks benefit from ample shade.) For maximum color, during the molt and for 6 to 8 weeks prior to a show, a good ration can be made by mixing four parts by volume game-bird flight conditioner, two parts oats, two parts cat kibbles, and one-half part black oil sunflower seeds.

It pays to go over a Cayuga with a tweezers to remove any white feathers prior to taking it to a show. Pay special attention to the small feathers of the head and neck. One white feather in a black duck can make the difference between a bird's winning first and not placing at all. (It is not permissible to remove large wing or tail feathers or to artificially color any feathers.)

Comments. For butchering, Cayugas must be plucked when they are in full feather to produce a clean-looking carcass. Pinfeathers can be eliminated altogether by skinning rather than plucking. In keeping with their ebony plumage, some Cayuga ducks produce their first eggs of the season encased in a black or gray cuticle. As the season progresses, egg color normally fades to light gray, blue, green, or even white.

Crested

Ducks adorned with feathery bonnets were depicted in art dating back more than 2,000 years. Some of the old Dutch painters included crested ducks of various colors in their rural scenes.

The crest is caused by a dominant mutation that has shown up with relative frequency throughout recorded history. Commonly, in any breed of Mallard descent, a duckling with a crest will hatch out of every 100,000 to 1,000,000 eggs hatched.

Today in some European countries, crested versions of most breeds are bred and shown. In North America the White Crested was included in the first *American Standard of Perfection* in 1874, with blacks being added in 1977. Even though Crested ducks are reasonably good layers and produce plump roasting birds, the majority are currently raised for pets and decoration. In many parts of North America, relatively few Cresteds are seen at poultry shows; because of their regal appearance, however, they always create considerable interest when they are exhibited. Demand for ducklings and breeding stock often exceeds supply.

Description. In all regards except for the crest, this is a normal-looking duck of medium size. The *American Standard of Perfection* states that the carriage is "nearly horizontal," but almost every Crested duck I have raised or seen has stood at 20 degrees or more above horizontal. The size, shape, and placement of the crest in day-old ducklings is an accurate guide to what the head adornment will be on the adult duck.

Varieties. The two officially APA-recognized varieties are the White and the Black. However, ducks with crests can be found in most colors or patterns. (See chapter 10 for color descriptions and information.)

Selecting breeders. First, consideration should be given to birds that display abundant vigor, walk normally, and show no neurological or physical deformities such as abnormal balance, severely kinked necks, roached backs, or wry tails. Moderately long, plump bodies with deep chests are desirable.

A good breeding trio of White Crested ducks, bred by the late Henry K. Miller. Breeders do not necessarily have to have show-type crests to produce some offspring with good crests.

To help produce a higher percentage of viable crested ducklings, the Wissenschaftlicher Geflugelhof Institute in Germany has come up with an ingenious test for selecting potential breeders. Each adult crested duck is placed on its back on a surface with good traction. Those that can right themselves consistently in the shortest length of time have been shown to produce more offspring and ones with fewer neurological abnormalities.

It is not necessary for breeders to have large crests for them to produce some ducklings with large headgear. Therefore, it is more important to select breeding birds that have firmly attached crests with smooth, rounded surfaces and no obvious divisions and that are centered as high on the skull as possible. No matter how good the parents are, some ducklings will be produced that have small, misshapen, lopsided, loosely attached, or totally absent crests.

Selecting and preparing show birds. Everything else being equal, the larger, rounder, and more firmly attached the crest, the more valuable the bird. Look for crests that are attached fairly high on the back of the skull. Ideally, the front edge of the crest is even with the eyes. Most judges prefer full-size birds (7 pounds [3 kg] for old drakes) with solid build, but excessive size and coarseness resembling Pekins should be avoided.

To maintain the crest's good condition during the show season, birds should be kept in a clean environment and the females (if necessary) separated from active drakes that may pull out feathers during mating. During the molt and for 6 to 8 weeks prior to a show, a good ration can be made by mixing four parts (by volume) game-bird flight conditioner, two parts oats, and two parts cat kibbles. Check crests for lice or mites, and treat as needed. Show cages used for Crested ducks must be a minimum of 27 inches (69 cm) tall so that the crest does not rub against the top of the pen.

Comments. Many people are attracted to the royal appearance of Crested ducks. However, these birds do have some inherent challenges. Crests in ducks (unlike geese and chickens) are caused by what is often described as an incompletely dominant autosomal semilethal allele, which has highly variable expressivity and, sometimes, incomplete penetrance. Theoretically, when two ducks with crests ($Cr^{Cr}Cr^+$) are mated together, approximately one quarter of their embryos (those homozygous for *crest* allele, $Cr^{Cr}Cr^{Cr}$) die in the shell or after hatching. Of those that survive, approximately two-thirds will be heterozygous for the *crest* allele ($Cr^{Cr}Cr^+$) and one-third will be homozygous for *wild type* (no crest, Cr^+Cr^+). In those ducks that carry the allele for *crest* ($Cr^{Cr}Cr^+$) the head adornment may range from being totally invisible to measuring 5 or more inches (13 cm) in diameter.

Due to the variability in the expressivity and penetrance of the *crest* allele, the above percentages can vary dramatically. We had matings that consistently produced significantly higher percentages of crested ducklings and lower numbers of nonviable embryos, suggesting that some homozygous crested individuals are viable.

Ducks that carry the *crest* allele are susceptible to neurological abnormalities (which may be apparent at hatching time or not until months later) and to skeletal deformities (kinked necks, shortened bodies, roached backs, wry tails). Fortunately, the majority of Crested ducks are normal and have a life expectancy similar to that of other breeds. Eggshell color can be white, tinted, blue, or green.

Orpington

Englishman William Cook, the famous poultry breeder and promoter who lived near Orpington in the southeast county of Kent, originated the Orpington breeds of both chicken and duck. Near the turn of the twentieth century, buff-colored plumage was highly prized in the British poultry world, and Buff Orpington ducks were developed to cash in on this fad.

Jonathan M. Thompson of Norfolk, England, has researched the origin and history of waterfowl breeds and uncovered information that challenges the frequently published accounts of the origin of the Orpington duck. In his paper *The Orpington Ducks (A cautionary tale)*, 2009, Mr. Thompson references articles in *The Poultry Journal* (1886–1904) that state the first Orpington Ducks were the Blues. They were introduced to the public in 1896, with the Buff coming out in the fall of 1897. Compared with most historical accounts, this reverses the order of varietal development and predates the time of origin by a decade. The foundation breeds are given as Aylesbury, Rouen, and Indian Runner. William Henry Cook, the son of William Cook, claims to have developed the Black and Chocolate varieties (both of which have white bibs), introducing them in about 1920, and includes the Cayuga as a foundation breed.

In England, where Orpingtons were promoted by the Orpington Duck Club, they performed well in the popular egg-laying contests (at the Harper Adams Agricultural College Laying Tests of 1928, Buff Orpingtons averaged 204 eggs in 12 months) and were raised in fair numbers as quick-maturing meat ducks.

Orpingtons made their appearance in the United States in 1908, when their originator showed a pair of Buffs at the prestigious Madison Square

Garden Show in New York City. In 1914 they were admitted to the *American Standard of Perfection* under the name of Buff Duck. This is an odd and confusing name that breaks standard nomenclature procedure for poultry (in no other instance is a color used as a breed name). Both for clarity and historical perspective, Orpington should be used to designate the breed and Buff the variety. Blue Orpingtons were advertised and shown occasionally in North America during the first half of the twentieth century but apparently were ultimately absorbed by the Swedish breed.

No one ever questioned the fine practical qualities of the Orpington, but they arrived in North America 35 years after the Pekin and were never able to compete with them as a commercial market duck. Currently, Buff Orpingtons are raised in modest numbers as an all-purpose duck, and at shows they are typically numerically second to Cayugas in the Mediumweight class.

Description. In conformation the Orpington is similar to the Cayuga. The significant differences between the two breeds are that the Orpington should have a slightly longer body and its longer bill should be attached slightly higher on the head with a nearly straight topline, reminiscent of its Runner heritage. Body carriage is approximately 20 degrees above horizontal.

A pair of Buff Orpingtons, bred by the late Henry K. Miller, with excellent bill and head shape, body conformation, and even plumage color.

Varieties. In North America Buff is the only color recognized by the American Poultry Association. (See chapter 10 for color descriptions and information.)

Selecting breeders. Many Orpingtons are significantly under standard weight and have short bills with excessively concave toplines. Therefore, full-size birds with long straight bills that are attached high on the head make valuable breeders. When possible, avoid birds with obvious white on their necks or elsewhere on the surface of their plumage.

Selecting and preparing show birds. The bird with the best plumage color often wins. Look for a medium shade of warm fawn-buff that is as uniform as possible in all sections and has a minimum of blue or gray shading. A white neck ring is a disqualification. Top show prospects also have good size, long bodies, deep chests that are nicely rounded, and nearly straight bills that are attached fairly high on the head.

During the molt and for 6 to 8 weeks prior to a show, you can use the same feeding recommendations as outlined for Crested to encourage excellent condition. Because buff-colored plumage is susceptible to fading, it is a common practice to keep ducks of this color out of direct sunlight from the onset of the molt through the end of the show season.

Comments. A group of well-bred Buff Orpingtons is a lovely sight, especially when they are foraging on a carpet of green grass or loafing on a blanket of snow. They do not show soiling as easily as white-plumaged ducks, and yet their pinfeathers are not unsightly when they are dressed for meat. Eggshell color is usually white or tinted. When out of a high-producing strain, Orpingtons can provide a good supply of eggs, medium-size roasting ducks, and beautiful show birds.

Swedish

Blue-colored ducks have been known in Europe for centuries. Tradition held that blue ducks were exceptionally hardy, superior meat producers, and difficult for predators to see. As early as 1835 the foundation stock of the Swedish duck was being raised by farmers in Pomerania, which at that time was part of the Swedish Kingdom but today straddles northeast Germany and northwest Poland.

Swedish ducks were imported into North America in 1884 and included in the *American Standard of Perfection* in 1904. Since then, they have been raised in modest numbers as general-purpose farm ducks and for pets, decoration, and exhibition.

A Blue Swedish drake with outstanding conformation and plumage markings bred by Ryan Gartman of Wisconsin. More important than the exact size of the bib are symmetry and clearly defined borders.

The original written standard called for the two outermost flight feathers on both wings to be white, with the remainder of the wing being blue. This unreasonable specification discouraged many people from showing this fine breed. In 1998 the standard was changed to allow either two or three of the outer flights to be white. This revision makes it slightly more likely to meet the standard specifications for wing color. However, both breeders and judges need to keep in mind that the color and pattern of the Swedish is one of the most challenging to perfect and that placing too much emphasis on such a minor detail as the exact number of white flights is a serious detriment to the breed.

Description. Compared with Cayugas and Orpingtons, Swedish are broader and not quite as long in the body, giving them a slightly stockier appearance. However, short bodies are not desirable in this breed, although it is a commonly seen fault. The head is oval-shaped and medium to medium-large in size. The medium-length bill should be nearly straight along its topline, as in the Orpington. But in the Swedish the bill should be attached to the head slightly lower, with a bit more rise to the forehead. Body carriage is approximately 20 degrees above horizontal.

Varieties. Blue is the only standardized color in North America, but Swedish also produce Black, Silver, and Splashed offspring. (See chapter 10 for color descriptions and information.)

Selecting breeders. Top priority should be given to vigorous, strong-legged birds with solid, well-muscled bodies of good size. Common faults to avoid include short, narrow, or shallow bodies; tails that are constantly cocked upward; narrow, snaky heads; and bills that are excessively long or have a distinctly concave topline.

Selecting and preparing show birds. Judges admire a Swedish with good size, adequate length, rich blue plumage with as little foreign color as possible (even the best specimens have a bit of black mixed with the blue), and a well-defined white bib. Although often not present, it is desirable that the feathers of the shoulders, back, and flanks be distinctly laced with a darker shade of blue.

During the molt and for 6 to 8 weeks prior to a show, you can use the same feed mix recommended for Crested. Although blue plumage is more stable than buff, over time sunlight will discolor it. Therefore, keeping blue ducks out of direct sunlight from the onset of the molt through the end of the show season will improve feather quality and clarity of color.

Comments. When well bred, the Swedish is a beautiful duck that rewards the person who has the patience to perfect both the color and pattern of this hardy breed. Eggshell color can be white, tinted, blue, or green.

7

Heavyweight Breeds

THE OLDER BREEDS IN THE HEAVYWEIGHT CLASS traditionally were raised primarily for meat. Typical weights range from 7 to 12 pounds (3 to 5.5 kg), although some individuals will tip the scales at considerably more. Characteristics that most of these breeds share include heavily muscled bodies, exceptionally fast growth, moderate rate of lay, the tendency to stay close to home if well fed, and calm temperaments.

Ducklings will reach their full size only if they are raised in the right environment and supplied an ample quantity of feed that meets all of their nutritional requirements. Inadequate nutrition can result in birds reaching as little as 60 percent of their genetic weight potential.

On the other hand, ducks that are going to be used for breeding stock or pets will live longer, reproduce better, and have fewer leg problems if they are not pushed for the fastest possible growth rate or maximum body weight.

Over time there is a tendency for large breeds to gradually lose size. Therefore, it is important to select breeding birds in each generation with good size; however, the very largest drakes are sometimes not fertile. A common practice is to select for breeding purposes the largest worthy females, mating them with active drakes that have good size but are not necessarily the biggest. When selecting potential breeders, be aware that occasionally there are birds that appear to be huge females but in fact are androgynous and will never lay eggs.

For good production, it is important that Heavyweight ducks are in good flesh but not obese as they enter the breeding season. Obesity can cause lameness, infertility, and prolapsed oviducts. In breeds that have profusely feathered abdomens (primarily Pekin and deep-keeled Rouen and Aylesbury), fertility is often improved if the feathers for 3 inches (7.5 cm) on all sides of the vent are trimmed short with scissors during the breeding season.

In single-male matings a vigorous one- or two-year-old drake that is not overweight and is consuming a breeder diet that encourages good fertility normally will service two to four ducks (in the very large, deep-keeled, exhibition-type Rouen and Aylesbury, pair or trio matings are usually the most successful). In flock matings a ratio of three to five ducks per drake normally gives good results. Some ducks in this class are good to excellent natural mothers if allowed to nest in a safe location. Extremely heavy ducks may get broody, but they have a tendency to crush their eggs or ducklings.

Appleyard

The large Silver Appleyard was developed by Reginald Appleyard in the 1930s at his famous Priory Waterfowl Farm near Bury St. Edmund, in West Suffolk County, England. In a farm brochure from the 1940s, Mr. Appleyard states that his purpose was to "make a beautiful breed of duck, with a combination of beauty, size, lots of big white eggs, and deep, long, wide breast."

Following World War II, interest in poultry breeds declined, and the Appleyard almost disappeared. Tom Bartlett of Folly Farm near Burton-on-the-Water, England, was instrumental in reviving this fine breed during the last third of the twentieth century.

Appleyards were brought to the United States in the late 1960s. However, it was not until 1984 that they were made available to the public and exhibited in Oregon. By 1988 Appleyards had gained sufficient popularity that at the Boston Poultry Show they constituted one of the larger classes of ducks. The qualifying meet for admission into the American Poultry Association (APA) *Standard of Perfection* was held October 1998 at the Minnesota State Poultry Association Show in Hutchinson, with 106 Appleyards shown by 12 exhibitors. Currently, Appleyards are being raised in increasing numbers for exhibition, pets, decoration, eggs, and gourmet roasting ducks.

Description. The Appleyard is a moderately large and sturdily built duck that should not have the extreme coarseness of features nor the loose feathering of the Pekin. The bill is of medium length with a nearly straight topline. Nearly oval with a slight forehead crown, the moderately large head is attached smoothly to a neck that is moderately thick, medium in length, and slightly arched forward. The compact body, which is carried at 15 to 25 degrees above horizontal, is broad and deep with a prominent chest that is smooth and well rounded.

In Great Britain Appleyards are bred and shown both plain-headed and with crests. Even though we have only bred from plain-headed birds for many generations, crested ducklings are still produced with fair regularity from some matings.

Varieties. Silver is the only recognized color variety, but pure white ducklings are sometimes produced. (See chapter 10 for color descriptions and information.)

Selecting breeders. Appleyards, along with Saxony, are the most active foragers and the best layers among the Heavyweight ducks, so it is important to use breeders that are robust, strong-legged, and excellent producers of large eggs. Many Appleyards are undersized, so birds with big, solid-muscled bodies are valuable. On the other hand, excessively large size (which currently is a rare problem) hampers foraging and laying ability and is not desired.

Selecting and preparing show birds. Big, solid birds with no keel on the breast, smooth silky plumage, and proper color will catch the judge's eye. If provided a clean environment, proper

A Silver Appleyard old drake, bred on our farm, with the desired plumage pattern. The silvery-white head markings typically become more pronounced with age.

A Silver Appleyard old duck, bred on our farm, with excellent conformation and plumage pattern. Female Appleyards tend to darken with age.

diet, and clean bathing water, Appleyards will usually be in good condition for showing. During the molt and for 6 to 8 weeks prior to a show, feeding a ration consisting of four parts (by volume) game-bird flight conditioner and one part cat kibbles will encourage excellent feather condition.

Comments. Compared with Pekins, Appleyards do not grow quite as fast, but they have much more interesting colors, are better layers, produce meat with more flavor and less fat, are better foragers, and are more likely to incubate and hatch their eggs. Most females lay white-shelled eggs. Appleyards are one of the best all-purpose large breeds of ducks and adapt to a wide range of environments.

Aylesbury

In the Vale of Aylesbury, less than 40 miles (65 km) northwest of London, Buckinghamshire "duckers" were renowned by 1800 for the production of England's finest white-skinned market ducklings. The famous pink-billed ducks of this region were originally known as White English, but by 1815 they were being called Aylesbury. Along with their unusual bill and skin color, these ducks were held in high esteem for their large size, ability to produce eggs in the winter months, fast growth, light bone, and a high percentage of mild-flavored meat when butchered at six to nine weeks of age.

For years it was thought that this unique duck could be raised successfully only in its home region, where there was an ample supply of a pumicelike white gravel found in the local streams. The common belief was that by digging their bills into this peculiar grit and consuming some of it, the birds' bills and flesh were lightened to the pale pink color so highly prized by the British epicures. Over time it was discovered that the Aylesbury could be raised in other locations, although not always with the same degree of pale color in the bill and flesh.

As poultry breeders began showing their prized fowl, some strains of Aylesbury were selected for greater size and deeper keels. Eventually two types evolved — production-bred market birds and standard-bred show specimens. At the first British poultry show held in 1845, two classes were provided for ducks: "Aylesbury or other white variety" and "Any other variety."

The Aylesbury was one of the first duck breeds to arrive in the United States from Europe. However, because yellow-skinned market birds have traditionally sold for higher prices in North America, the commercial use of Aylesbury never flourished in Canada or the United States. Furthermore, nineteenth-century American duck growers considered the Aylesbury to be

An eight-month-old Aylesbury duck with great body depth and excellent back profile (topline). Bred by Good Shepherd Ranch in Kansas.

less hardy than other breeds. In retrospect this perceived lack of hardiness was likely caused by diets that were deficient in vitamins A and D₃.

Aylesburies were exhibited at the inaugural poultry show in America, held in 1849 in Boston, Massachusetts. In 1874 they were included in the first *American Standard of Perfection*. Although not exhibited in large numbers at most shows, Aylesburies have had a devoted following over the decades among specialty breeders and continue to win their share of honors.

Description. With a long body, horizontal carriage, and pendulous keel that runs from stem to stern and brushes the ground in top specimens, the Aylesbury is a bona fide low rider. Ideally, the topline of the back is only slightly convex from the shoulders to the tail. However, many Aylesburies, especially those that are extra large and long-bodied, have a fair amount of arch to their backs when they are standing still and relaxed. The unique head has been compared to that of the American woodcock, with a flattened forehead, high-set eyes, and a long, straight bill that ideally measures 3 to 3½ inches (7.5 to 9 cm) or more in length. The neck is moderately long with a slight forward arch when the bird is relaxed.

The plumage is the whitest of all ducks, the feet and legs are orange, and the eyes are dark grayish blue. The natural color of the bill is pink in day-old ducklings and pale yellow to blanched yellowish-pink in nonlaying adults.

After they have been laying for a time, bill color in most females will fade to very pale pink, usually with black streaks or spots in the bean.

Varieties. Aylesbury are bred in White only.

Selecting breeders. The Aylesbury is a complex breed that is a far departure in size and conformation from the wild Mallard. It is one of the more challenging breeds to perfect. Ideal breeders will seldom — if ever — be found, so putting together a mating is a balancing act. The rule of thumb is this: Try to avoid mating two birds with the same obvious faults.

It is critical that breeders have strong legs, display good vigor, and be free of genetic physical deformities. Extreme length and depth of body are desired, along with good width of back. However, the largest males with the most exaggerated type are sometimes not fertile, so it can be risky to rely on them as breeders.

Common faults to watch for include small size; bodies that are short or shallow; high-crowned foreheads; bills that are short and/or distinctly concave along the topline; body carriage that is obviously elevated in front; wings that slide far down the sides of the body or are constantly held excessively high with the tips crossed over the back; and keels that are shallow, short, wry, or uneven along the bottom line or do not have a prominently curved bow on the front.

When possible, avoid using drakes that have black pigment anywhere on their bills, especially if they are two years old or younger (keep in mind that injuries to the bill sometimes cause transitory discoloration). Be aware that if excessive effort is made to eliminate all black from the bills of sexually mature females, the productivity of the strain will usually decline.

Selecting and preparing show birds. A good show specimen has a profile resembling a rectangular cement block with rounded edges; a long, straight, whitish pink bill; and satin white plumage. To obtain the ideal bill and plumage color, for at least 6 to 8 weeks prior to showing, Aylesburies need to be kept in a clean environment with nonstaining bedding, provided clean bathing water, kept out of sunlight, and fed a diet that is low in yellow pigments. Feed ingredients that help promote pure white plumage and pale pink bills include oats, barley, white wheat, white rice, roasted soybeans or soybean meal, fish meal, meat meal, dried milk, and distillers' solubles. A ration that promotes good health and encourages decent feather and bill color consists of four parts (by volume) game-bird flight conditioner, two parts oats, two parts white wheat, and two parts fish-based cat kibbles.

For ducks that are kept in shade and consume a diet low in yellow-pigmented ingredients, it is critical that a good multivitamin poultry supplement be given for birds to maintain good health and physical appearance. The amount of vitamin supplement required depends on the vitamin levels in the regular diet. In general, if the vitamin supplement is given in an amount that provides a daily dosage of 100 to 200 ICUs (international chick units) of vitamin D_3 and approximately 1,000 IUs (international units) of vitamin A per bird, the concentration of the other vitamins in the supplement will be appropriate. Note: Birds do not utilize the D_2 form of Vitamin D efficiently.

Comments. The unique pink bill and pink/white skin color of the Aylesbury results from the interaction of genetics, diet, and environment (see Bill, Skin, and Leg Color, chapter 10). White-skinned ducks have elevated dietary requirements for vitamins A and D_3, especially during the breeding season, during times of low sunlight intensity, and when succulent, fast-growing greens high in xanthophylls are not readily available.

Fertility is often improved if swimming water at least 6 inches (15 cm) deep is available. Keep in mind that Aylesburies that are excessively fat rarely reproduce well. We had good reproduction with large, exhibition-type Aylesburies by feeding them a high-quality waterfowl breeder ration in limited quantities that they would clean up in 10 to 15 minutes at morning and evening feedings. We started this feeding regime one month prior to and continued it throughout the breeding season. Females typically lay white, tinted, or green eggs. Aylesburies are easily tamed and make responsive and unique pets.

Muscovy

The wild Muscovy (*Cairina moschata*) is a nonmigratory native of Mexico, Central America, and much of South America east of the Andes Mountains. Being a tree duck, they roost in trees at night and spend hours preening on elevated perches after their daily bath. Preferred nesting sites are on large branches and cavities in tree trunks.

Muscovies are one of the larger wild ducks; drakes average 4¼ to 6 pounds (2 to 2.75 kg) and ducks 2¼ to 3 pounds (1 to 1.5 kg). Both sexes have blackish plumage with metallic green sheen. Juveniles, especially, have dark brown or bronze on the chest and underbody. The brilliant white wing badges develop gradually with age, being the most visible during preening and in flight, when they flash unmistakably. The area around the eyes of the sexually mature drake is covered with smooth, black skin that is jeweled with one or more red spots. The junction of the forehead and bill is adorned with a black, fleshy nub that is highlighted with bright red. Females are nearly devoid of facial skin patches, with only a rudimentary nub on the base of the bill. In both sexes the feet and legs are dusky black. The bill is black, with a bluish white or pinkish white band across the end, then tipped with a blackish bean.

By the time Columbus arrived in the New World, Muscovies had most likely been raised in domestication for centuries. In 1514 Spanish conquistadors found the native people raising several colors of Muscovies on the northern coast of what we now know as Colombia. From their home in the Americas, Muscovies probably traveled to Africa and then to Europe, where they arrived by at least the mid-1500s.

A host of theories have been advanced as to how this native American became known by a Russian name. A reasonable suggestion that is logistically possible maintains that these birds were transported to England by the Muscovite Company during the 1500s. Since it was a common practice to attach the importer's name to products it traded, it would have been an easy transition from Muscovite Duck to Muscovy.

A Blue Muscovy yearling duck, bred on our farm, with the conformation, textured facial caruncling, and dark lacing desired on the body plumage of show specimens.

In North America the Muscovy has a long history, especially in Mexico, of being raised as a self-reliant duck that hatches and often raises multiple broods a year with a minimum of assistance from humans. However, they did not become a major market duck in Canada or the United States until the latter part of the twentieth century. Today, as the demand for leaner meat and liver pâté steadily increases, the breeding and marketing of Muscovies and their hybrids are expanding.

When the first American poultry show was staged in Boston in 1849, three people exhibited Muscovies. In 1874 when the first *American Standard of Perfection* was compiled, the White Muscovy was included, even though colored birds were more common. Exhibition Muscovies have experienced a renaissance in both quantity and quality in the United States and Canada since the early 1990s.

Description. It takes only a glance to realize that domestic Muscovies are not only a unique breed but also a different species from other domestic ducks. A closer look reveals the physical characteristics that allow its wild ancestors to be at home in tropical rainforests.

The medium-length bill is relatively narrow, with large nostrils, a fairly concave topline, and an extra-large nail (bean) at the tip. The head of the mature drake is massive, and its face is covered with red skin. The duck has a medium-size head and smaller facial skin patches. Both sexes can raise the feathers on top of their heads into a crest when excited, flirting, or angry. The male's crest feathers are longer and stylishly waved.

Compared with other domestic ducks, the body is flattened, heavily muscled, and extra wide across the shoulders. The wings are very wide and moderately long, with the tips being more rounded than in other breeds. The tail is long and broad. (The wings and tail were designed for flying through heavily forested terrain.) The long toes are webbed, amazingly strong, and tipped with talonlike claws.

The size differential between the sexes is remarkable. At hatching, Muscovy ducklings are the same size. By three weeks of age, females weigh approximately 65 percent as much as the males, at twelve weeks 55 percent, and at twenty-six weeks 45 to 50 percent. For comparison, in other domestic breeds and wild species that nest on the ground, ducks typically weigh only 8 to 12 percent less than drakes at maturity. In species that nest in tree cavities, such as the American Wood Duck, Bufflehead, and Common Merganser, females on average weigh 23 to 31 percent less than males. It is possible that in tree-nesting species, smaller females are more likely to find safe nesting cavities.

As Muscovies go about their daily activities, they frequently wag their tails from side to side and thrust their heads front to back. A less frequent but still common display includes pointing the bill to the sky and rapidly moving the mandibles as if catching invisible raindrops.

When Muscovies are on uncrowded free range, there is usually a minimum of serious fighting. However, they can battle with amazing fervor if their territory is too small. Sometimes they face each other like fighting chickens, leaping into the air and striking at one another with their claws. Other times they fight by pecking each other with their bills and pounding with the bony protuberance on the leading edge of their wings. When breeding pens of Muscovies are adjacent to one another, it is usually beneficial to have solid metal or wooden dividers at least 2 feet (0.6 m) high along the entire length of the pens so the drakes in adjoining pens cannot see each other and attempt to fight through the fence.

Muscovy ducklings can be identified from other breeds by their medium-size bill with a particularly large bean on the tip, flattened bodies, distinctive trilling vocalization, larger and stiffer tails, and needle-sharp claws. (When naturally hatched in tree cavities, the ducklings use their stiff tails and sharp claws for climbing out.)

Varieties. White, Black, Blue, and Chocolate are recognized by the APA. Wild Type, Silver, Lavender (sometimes called Self Blue), Buff, Blue Fawn, Lilac, Pastel, Barred, Rippled, White-headed, Magpie, Splashed, and others are also raised. (See chapter 10 for color descriptions and information.)

Selecting breeders. In order to choose breeding stock wisely, you need to have clearly in mind your main purpose for raising Muscovies. Remember, there are various distinct strains, and characteristics such as size, growth rate, feed efficiency, egg production, maternal instincts, and foraging ability can vary drastically among them.

If your goal is to have self-reliant birds that will find much of their own food and hatch and raise their own babies, the main criteria you should consider in making your selection should include vigor, foraging ability, and maternal instincts. Birds that are medium- to dark-colored have better camouflage in most settings. Medium-size birds that can fly are typically better foragers and are also less likely to be caught by predators. Many commercial strains have had their maternal instincts bred out, making them worthless as natural mothers.

A MEMORABLE MOMENT

To be on that river deep in the interior of Guatemala was to experience a real-life fantasy — miles of meandering milky green water; dozens of hand-size, electric blue butterflies; and a towering ceiba tree with a pair of toucans hopping along a massive branch. As our canoe rounded the bend, the sleek, black duck vaulted from its perch 15 feet up in the snag jutting out of the river and flew straight into the jungle. Our first sighting of a wild Muscovy! From that brief encounter, it was concluded that we had just crossed paths with a wily, old bird. The first hint of its veteran status was the bold white forewing patches that only older birds display. Second, rather than taking the open flyway down the river, which would have made it an easy target for the many hunters who plied these waters, it vanished straight into the woodland. After having seen domestic Muscovies in nearly every village and marketplace we had visited, it was a thrill finally to glimpse one of the untamed originals.

If your goal is to produce the maximum number of market birds, then the primary selection criteria should include good egg production, fast growth, efficient feed conversion, heavy breast muscling, and a high meat-to-fat ratio. For many markets birds with good size are desirable. However, as size increases beyond a reasonable level, it is normal for productivity, feed efficiency, and growth rate to decrease. Huge show birds usually do not make efficient producers of market ducks.

For the production of show birds, the main characteristics that you should select for include massive size, good type, proper plumage color, and well-caruncled faces with a minimum of black (caruncles are the bumpy skin on the face). If the Muscovies are descendants of a strain with good faces, it is not necessary for breeding females to have exceptionally large facial skin patches to produce offspring with good exhibition faces. In fact, it is not unusual for females with male-like caruncling to be sterile due to a hormonal imbalance.

Selecting and preparing show birds. Good show specimens are massive throughout, with long bodies and wide shoulders. The head is wide across the skull, and in mature birds the face is covered with well-defined caruncles that are grainy right up to the eyes. In old drakes a fleshy knob the size of a grape or larger lies above the nostrils on the upper mandible. The knob on hormonally normal females is much smaller. The caruncling should be balanced on both sides of the head. It is fine for the caruncling to extend down the neck in old drakes, but it is a serious fault if it blots out the crest feathers on top of the head or makes it impossible for the bird to see.

To have Muscovies in top condition for showing, provide them with a well-drained, clean pen with mounds of clean straw for them to perch and preen on; clean bathing water (dirty water and mud can ruin a White Muscovy for showing); plenty of shade (although some direct sunlight is needed to get the brightest red color in the face); a daily feeding of succulent greens such as lettuce, chard, and tender grass clippings; and a ration consisting of two parts (by volume) game-bird flight conditioner, one part oats, and one part cat kibbles, plus insoluble grit. Once they develop a taste for it, Muscovies greatly enjoy a teaspoon per bird of canned cat food several times a week, and it seems to improve feather quality. To show off to their best advantage, large drakes need to be penned in double coops at shows.

Comments. Muscovies are often described as mute. Although this is not literally accurate, they are by far the quietest of all breeds. Females occasionally quack weakly but mostly use a variety of soft chirps and whistles to communicate. The drake's primary vocalizations are hoarse, breathy exhales of varying lengths.

Muscovies can appear to have extremely laid-back personalities. When walking through their territory, you can almost step on them before they move. But as soon as they know your intent is to catch them, they are uncannily quick and adept at avoiding your grasp. Because they do not herd as well as other ducks, a good way to catch them is to walk them into a V-shaped corner one at a time, then snatch them with your hands or a long-handled fishnet. Hand-tamed pets can often simply be picked up off the ground.

Muscovies are by far the strongest of all ducks and with their powerful legs and long claws are capable of inflicting painful scratches on your hands and arms when they are caught. I wear long-sleeved denim coveralls and leather gloves when handling them.

Some people extol the Muscovy as being the sweetest-tempered of all ducks, while others insist that they are nothing short of the fabled Tasmanian Devil incarnated in duck form. Depending on the particular bird and the circumstances, both descriptions are reasonably accurate. Some drakes live amicably for years in yards with all manner of fowl and animals, whereas others will mercilessly drag down and attempt to mate with critters of many shapes and sizes. Full brothers can have personalities that are polar opposites. Some females, particularly when new birds are introduced to their territory, will attack like gamecocks. When serious assaults persist for more than a brief time, either the victim or the attacker should be removed. While I have found no failproof way to predict personality, older individuals and birds that have been raised in small groups tend to be the most aggressive. Conversely, those that have been raised around diverse creatures and in larger groups typically are more tolerant.

Most females, if not overly fat, are fair to good flyers and will perch on fences and buildings. If they have sufficient food and feel safe, they normally will not leave their home domain. The annual clipping of the 10 primary flight feathers of one wing prevents sustained flight. Most mature drakes are too heavy to become airborne for more than a short distance. However, if sufficiently motivated, they are capable of climbing over 3- to 4-foot-(1 to 1.25 m) high wire fences.

Muscovies enjoy swimming and if permitted to do so will bathe daily. However, their feathers do lose water repellency more easily than most ducks. Individuals with poor feather condition or those unaccustomed to swimming can become waterlogged if forced to stay on water.

Their tropical roots notwithstanding, Muscovies can tolerate considerable freezing temperatures as long as they are protected from cold winds and have a draft-free, well-bedded shelter to sleep in at night. In wet weather they may need access to a roofed, dry area.

Muscovies are highly prized for their meat and are easier to pluck than other ducks. They have exceptionally broad, well-muscled breasts and are one of the leanest of all waterfowl. We have found that their meat makes a fine "beef" stew, while cured and smoked Muscovy is similar to lean ham.

Muscovy eggs can be hatched in incubators, but for good success incubation temperature and humidity must be unusually precise. In my experience the best results are obtained when Muscovy eggs are incubated in a separate incubator; the temperatures and humidity levels are recorded each day; and from the 15th to the 32nd day of incubation, the eggs are cooled once daily by spraying them with lukewarm water. At the end of the incubation period, the daily temperature and humidity readings are added up and then divided by the number of days. If the hatch was a success, the incubator should be operated at the same average daily temperature and humidity levels for the next batch of eggs. If the eggs hatched poorly, then the temperature and/or humidity should be adjusted.

The ducklings of this species have a reputation for being difficult to ship. However, we have had good success shipping them across the continent. The keys to shipping Muscovy hatchlings are proper incubation; mailing them no more than 12 to 36 hours after they emerge from the shell; and sending them only to destinations that they will reach in 48 hours or less.

Females are usually dedicated mothers, and it is not unusual to see them with 10 to 16 ducklings in tow. In fact, Muscovies are such reliable broodies that they are often used to hatch the eggs of other waterfowl, including rare species of wild ducks, geese, and swans. Their eggs are typically white to tan and have an incubation period of 33 to 35 days. Muscovy drakes normally fertilize more ducks than other Heavyweight males. In single-male matings a drake can usually fertilize 5 to 7 ducks. In flock matings a ratio of 6 to 10 ducks per drake usually gives good results.

As long as they are not overly inbred or out of a commercial strain, Muscovies are the most self-reliant of all poultry species in many situations. They are active foragers, consume more grass than other ducks, and are first-rate fly catchers. While ducks of most breeds will occasionally eat tiny rodents when the opportunity presents itself, Muscovies have greater carnivorous tendencies.

Pekin

The ancient people of China were among the first to domesticate the Mallard, probably before 1000 AD, and developed ducks into an important food source. Over time the Chinese became some of the world's most advanced duck raisers, developing sophisticated artificial incubation techniques and producing various types of ducks, including the Pekin.

The arrival of the Pekin in the United States is described in a letter written by James E. Palmer that was published in the September 1874 issue of the *Poultry World* and quoted by John H. Robinson in his book *Ducks and Geese for Profit and Pleasure* (1924). In his account Palmer states that he was asked by an American businessman, a Mr. McGrath, to transport 15 large white ducks from Shanghai to the United States. The ducks had been hatched from eggs acquired in Peking (today known as Beijing). Three drakes and six ducks survived the 124-day oceanic trip, landing in New York City in March 1873.

In 1874 the newly formed American Poultry Association published its first book of standards and included the Pekin. Within a short time after its introduction to North America, the Pekin became the primary breed raised for the production of market ducklings. The breed's popularity continues today, with

Pekins being raised in greater numbers than all other breeds combined. Their commercial success is due to a combination of traits that include hardiness; large size; unsurpassed growth rate (up to 8.5 pounds [3.75 kg] in 7 weeks); high feed efficiency (2.6 to 2.8 pounds [1.2 to 1.3 kg] of feed per pound [0.5 kg] of body weight); high fertility; excellent hatchability; calm temperament; and white plumage.

As show birds Pekins are popular for 4-H and Future Farmers of America projects, especially for market classes. In open-class competition at large shows, they are common but often not as numerous as Rouen or Muscovy.

Description. From early illustrations and written accounts, it is clear that the average Pekin raised in North America today has not changed significantly in appearance. The classic American Pekin has a thickset body that is carried at a jaunty 35 to 45 degrees above horizontal. The underbody is wide, smooth, and keel-less, with an ample abdomen that is profusely feathered. Unlike that of other breeds, the tail habitually sticks up above the line of the back. The head is large with full cheeks, and especially in females, the forehead feathers rise abruptly from the base of the bill to form an elevated crown. The neck is thick and moderately long and often appears to be arched forward (mostly

These huge seven-month-old, standard-bred Pekins weigh 12 pounds [5.5 kg] (the drake at left) and 11 pounds [5 kg] (the duck at right). The drake won Reserve Champion Heavy Duck at the 1999 Northwest Winter Classic. Bred by Gene Bunting.

because of the way the bill, head, and neck flow from one to the other). The bill should be of medium length, thick enough to be in harmony with the massive bird, and nearly straight along the topline.

In England and Germany, where the Pekin was introduced in 1874, Pekins stand almost vertical and have short, plump bodies; short, thick necks; chubby heads; and short bills. German Pekins were imported to North America in the 1990s, and their offspring, both pure and those crossed with American-strain birds, have been exhibited at some shows.

Varieties. Pekins are bred in creamy White only.

Selecting breeders. When Pekins first arrived in North America, the main selection criteria were vigor, strong legs, large size, and fast growth rate. Later, females were housed individually in narrow runs, and those that were the best layers were kept for breeding. Then interest grew in producing strains that were the most efficient at converting feed into body mass. While commercial breeders continue to select for all these traits, there is a growing emphasis today on developing market Pekins with less fat and more breast meat. The result is that a number of distinct types of Pekins are currently available, each with unique market traits. When acquiring breeding stock, choose the strain(s) that best fit your needs. In general, the largest and fastest-growing strains have the most body fat, whereas those that are the leanest grow a bit slower and consume more feed per pound of bird.

Pekins that have been bred for exhibition are larger and more massive than commercial strains. In this breed individuals that are good show birds normally also produce good exhibition offspring. Because they are large, always make sure to choose breeders that have stout legs and walk normally.

Selecting and preparing show birds. Top show specimens stand at approximately 40 degrees above horizontal and are massive in all sections. They have long, thick bodies; barrel-shaped chests; and powerful heads and necks (the exhibition Pekin has the largest head and neck of all breeds). To add style and mass to the head and neck, it is desirable for the feathers of the back of the head and upper neck to project back or slightly upward, forming a mane.

The Pekin is the only white duck that should have a strong cream or yellow cast to its plumage. The creamy hue is caused by ingested yellow and orange pigments. Therefore, diets rich in green feeds, yellow corn, marigold petals, and other yellow-pigmented foodstuffs will increase a bird's depth of color, especially if consumed 6 to 8 weeks prior to and throughout the molt and regrowth of feathers. A good show ration to enhance color and body size consists of four parts (by volume) game-bird flight conditioner, three parts yellow corn, two parts cat kibbles, and one part rabbit or alfalfa pellets. Yellow pigment in the plumage bleaches easily. The yellowest Pekins I have seen were exhibited by a poultry judge whose property had red clay soil.

Comments. Pekins are ideally suited for situations in which the quickest-maturing meat ducks are desired. They make good pets and are often kept on ponds "just for pretty." Females are talkative and lay white or tinted eggs.

Rouen

Among the tame ducks raised by French farmers several hundred years ago were some resembling large Mallards. Around 1800 these ducks reached England, where they were variously called Rhône (an area in southwest central France), Rohan (a Catholic cardinal), Roan (a mixture of colors), and Rouen (a town in north central France). Eventually, the name Rouen was settled on in both England and France.

With their amazing proclivity for producing and refining breeds of poultry, the British soon redesigned the Rouen. They altered its sleek "puddle duck" form into a thickset boat shape, doubled its size, and improved its colors. The Rouen was used primarily as a high-quality roasting duck when butchered at five to six months of age.

According to Paul Ives in his book *Domestic Geese and Ducks* (1947), Rouens were imported into the United States by D. W. Lincoln of Worcester, Massachusetts, in 1850. In North America the Rouen became popular as a colorful general-purpose farm duck and prior to the arrival of the Pekin was used for the production of market ducklings.

The Rouen was included in the first APA *Standard of Perfection* in 1874 and ever since has been regarded by many as the ultimate exhibition duck for its beauty, its size, and the challenge involved in breeding truly good show specimens. At major shows it often accrues the most entries among the Heavyweight class and is a frequent winner in competition with other breeds.

Description. There are two distinct types of ducks raised in North America that are known as Rouens. The common or production-bred Rouen is a Mallard-colored bird with "normal" duck conformation, generally weighing 6 to 8 pounds (2.75 to 3.5 kg). These are raised primarily as general-purpose ducks and for decoration.

The standard-bred Rouen is a huge block-shaped duck weighing 9 to 12 pounds (4 to 5.5 kg) with the most highly perfected Mallard coloration of any breed. These birds are bred principally for exhibition and decoration, but they make great pets and large, tasty roasting ducks. In size and type they are almost identical to the standard-bred Aylesbury, except for the shape of the head, bill, and back. In Rouens, ideally the head is rounder; the bill is a bit shorter and slightly more concave along the topline; and the back shows more arch from the shoulders to the tail.

Varieties. Gray is the original and only variety recognized by the American Poultry Association. Black, Blue Fawn, and Pastel varieties have been developed and occasionally exhibited. (See chapter 10 for color descriptions and information.)

Selecting breeders. In production-bred strains the selection criteria should be based on the characteristics that best fit your needs.

In the standard-bred Rouen, emphasis should be placed on vigor, strong legs, large size, good body length, deep keels that are straight and level, horizontal body carriage, and proper color and markings. Sometimes the largest and deepest-keeled drakes have poor fertility, so it can pay to retain for breeding some males that are not quite as extreme in size and keel development. The arch in the back of a Rouen normally becomes more pronounced with age, so one must be cautious in breeding from young birds with excessively arched backs.

A massive, standard-bred Rouen old duck bred by Danny Padgett of Florida and owned by Gene Bunting of Oregon.

Selecting and preparing show birds. In good competition the winners have excellent size, long bodies displaying good depth, straight keels that run from the breast to the abdomen, smoothly arched backs, fine color and markings, and sparkling condition. To get them in good show condition, Rouens should be kept in clean pens with fresh bathing water and sufficient room for exercise (I have used pens at least 10 feet [3 m] long). Rouens should be fed either free choice or twice daily a quantity that they clean up in 10 to 20 minutes for 6 to 8 weeks prior to and throughout the show season, with a ration consisting of three parts (by volume) game-bird flight conditioner, one part cat kibbles, and one part oat pellets or whole heavy oats to help promote keel development and fine feather condition.

Comments. Keep in mind that deep-keeled Rouens that are excessively fat rarely reproduce well. We had good reproduction with large, exhibition-type Rouens by feeding them a high-quality waterfowl breeder ration in limited quantities that they would clean up in 10 to 15 minutes at morning and evening feedings. We started this feeding regime 1 month prior to the breeding season and continued it throughout the breeding period. With deep-keeled Rouens, fertility is usually improved if swimming water at least 6 inches (15 cm) deep is available. Exhibition Rouens are so heavy that they tend to crush their eggs if allowed to incubate them, but some production-bred females are good setters. Egg color is usually white, tinted, green, or blue.

Saxony

In eastern Germany, about 25 miles from the Czechoslovakian border, Albert Franz of Chemnitz began working in 1930 on creating a new multipurpose duck. He used Rouen, German Pekin, and Blue Pomeranian in his breeding program. In 1934 his creation was introduced at the Saxony Show. Unfortunately, few Saxony remained at the end of World War II, so Franz renewed his breeding program. The Saxony was recognized as an official breed in 1957 in Germany, where it continues to be esteemed for its enchanting beauty and the production of large eggs and full-breasted roasting ducks of superior quality.

The Saxony was introduced in the United States in 1984. Their numbers have been increasing as they gain recognition as a hardy breed that combines unique plumage color with excellent practical qualities. The qualifying meet for admission into the American Poultry Association (APA) *Standard of Perfection* was held in October 2000 at the Western Waterfowl Exposition in Albany, Oregon, with 64 Saxony shown by 13 exhibitors.

Description. In type and size the Saxony resembles the Appleyard. It is a moderately large and sturdily built duck that does not have the extreme coarseness of features or the loose feathering of the Pekin. Due to its tight plumage and thick muscling, it is often heavier than it appears.

Its bill is stout, of medium length, and with a nearly straight topline. The moderately large head is fairly oval in shape, with only a slight forehead crown. The head blends smoothly into a medium-length neck that is moderately thick and slightly arched forward. The compact body is moderately long, broad across the shoulders, with a prominent chest that is smoothly rounded. The underbody is smooth and the abdomen moderately full. The back has good width and is nearly straight along its topline, with gently rounded shoulders. The tail is carried slightly above the line of the back. Body carriage is approximately 25 degrees above horizontal when the bird is relaxed but may go up to 40 degrees when excited.

Varieties. The name Saxony refers to both the breed and its single color variety. (See chapter 10 for color descriptions and information.)

A Saxony young drake (bred on our farm) with ideal conformation and plumage pattern. Note especially the clean head and neck color, and the white ring completely encircling his neck.

A Saxony old duck (bred on our farm) with the desired conformation and plumage pattern. Note especially the bold creamy white facial and neck front markings.

Selecting breeders. Along with the Appleyard, the Saxony is the most active forager and best layer among the large ducks of Mallard descent. Therefore, it is important to use breeders that are robust, strong-legged, active, and prolific layers of large eggs. Many Saxony are undersized, so heavily muscled birds of good size are valuable for breeding. On the other hand, excessively large size is not desired because it hampers foraging ability and laying efficiency.

Selecting and preparing show birds. The Saxony is a beautiful exhibition bird, and its unique colors provide a pleasing contrast to other breeds in the showroom. Moderately large, solidly built, no keel on the breast, silky smooth plumage, rich colors, and clearly defined markings are the main characteristics to look for in show birds.

The soft colors of the Saxony will fade and discolor under prolonged exposure to direct sunlight, so the birds should have access to ample shade. A diet consisting of three parts (by volume) game-bird flight conditioner, one part oat pellets or whole heavy oats, and one part cat kibbles will promote well-fleshed bodies and fine feather condition.

Comments. Compared with Pekins, Saxony do not grow quite as fast, but they have much more interesting plumage, produce meat with more flavor and less fat, are better foragers, and are more likely to incubate and hatch their eggs. Most females lay large, white eggs. Saxony are one of the best large all-purpose breeds of ducks and adapt well to a wide range of environments.

8

The Importance of Preserving Rare Breeds

FOR THOUSANDS OF YEARS people throughout the world have raised animals for companionship, work, food, clothing, pest control, beauty, and sport. Sometimes by chance but often by design, through the generations animals were selected that were well suited for specific tasks or environments. Soon after the turn of the twentieth century, the number of breeds of domesticated animals approached its zenith. Farm animals were suited to an amazing array of environments and uses.

As early as the 1920s, a new trend began to emerge. The mechanization and specialization of farms ushered in an era when fewer and fewer breeds were relied on to produce food. The livestock and poultry industries focused increasingly on specialized breeds and hybrids that could be packed tightly in minimal space. Improved yields and greater efficiency became the rallying cry of the "modern" agriculturists. With uncanny speed this tunnel vision led to near decimation, or in some cases total loss, of various old breeds.

By the 1960s whispers of warning could be heard. Some stubborn farm folks who had turned a deaf ear to the "experts" insisted that the old breeds still had their place in the modern world. In the halls of academia, a few forward-thinking geneticists pointed out that the losses of genetic stocks of plants and animals, both wild and domestic, could have profoundly negative effects on the quality of life on planet Earth, even decreasing humans' chances for long-term survival.

The Dutch Hook Bill has been traced to the 1600s; it nearly became extinct before a few breeders in Holland, Germany, and England saved the breed in the latter part of the twentieth century. This ten-week-old pair was part of the first documented clutch of Hook Bill ducklings hatched in North America. This fascinating breed, which we imported in spring 2000 and are breeding and studying at our Waterfowl Farm, comes in white-bibbed (shown here), nonbibbed, and pure white.

How You Can Help

As we head into the twenty-first century, preserving the widest possible gene pool in all types of plants and animals is essential. Around the world, geneticists, trained plant and animal breeders, and — just as important — thousands of lay people are dedicated guardians of endangered genetic stocks of all kinds. A variety of organizations now promote this work, including Seed Savers and National Germplasm Repositories (for plants) and the American Livestock Breeds Conservancy, the Society for the Preservation of Poultry Antiquities, and the International Registry of Poultry Genetic Stocks (for animals and fowl). Contact information for some of these is in appendix H on page 341.

Keep Rare Breeds

One of the most important contributions you can make is to raise one or more of the rare breeds. By purchasing endangered breeds you make three important contributions: You support the farms that breed them, you increase the populations of the breeds, and you augment the number of locations where the breeds are kept, thus reducing the possibility that they could be wiped out by plague, marauding predators, or natural disasters. Some people are concerned that if they buy rare breeds and then butcher them they are hurting the preservation efforts. Actually, they are helping by increasing demand for the breed and utilizing them for one of their original purposes.

Multiply Rare Breeds

A step beyond simply keeping rare ducks is breeding them. By propagating them you are increasing their numbers and providing yourself with additional birds from which to select future breeding stock. And you can distribute birds to other interested persons.

If you raise a sufficient number of young birds each year, you can retain the best specimens for your own breeding flock or distribute them to other people, and the less typical specimens can be used for food, pets, or decoration. If you save the superior specimens for breeding each year, you will maintain the overall quality of the gene pool or, better yet, improve it.

Learn about Your Chosen Breeds

Learning the history, conformation, color, and other characteristics of chosen breeds can increase your enjoyment, improve your ability to select breeding stock, and make you a better rare-breeds promoter. When promoting rare breeds, remember that people are usually attracted to a particular breed for one or more of the following reasons: appearance, practical qualities, or history. Experience has shown that the more one is engaged with the story of a breed, the more likely one is to continue raising it.

Exhibit Your Rare Breeds

One of the best ways to introduce people to heritage breeds and create interest is to exhibit your rare birds at county, state, provincial, or poultry club shows. When you are showing rare breeds or varieties, a wonderful way to educate people is to prepare a poster or card that gives pertinent breed information and place it on or near the display cage of the birds. An example of the type of information card we have used is shown at the top of page 103.

If printed on paper, this card can be affixed to a slightly larger piece of ¼-inch plywood, outfitted with two opened paper clips, and hung on a cage. You should ask the show superintendent for permission — in our experience

---

ANCONA DUCKS

Origin: Great Britain
Weight: 5 to 7 pounds (2.25 to 3 kg)
Color Varieties: Black, Blue, Chocolate, Lavender, and Silver.
Practical Qualities: Excellent layers (210 to 280 per year) of white, cream, green, blue, or spotted eggs; high-quality meat; excellent foragers that eat large quantities of slugs, snails, and insect pests; easy to see in field or on pond
Characteristics: The plumage pattern is broken with irregular white spots and patches throughout; seldom are two individuals marked exactly alike; the bill and feet are spotted
Status: Extremely rare
For more information contact: *(your name, address, phone number)*

Exhibit information card

superintendents have always been pleased with these information plaques because they add class and interest to a show.

An additional way to disseminate information at shows is to make a simple wooden holder for free flyers that describe your efforts (or others') to improve and preserve rare breeds. The holder with flyers can be placed in a prominent location, such as on top of the show cage.

Some rare breeds, because of their scarcity, have not been officially recognized by the American Poultry Association (APA), even if they are recognized in their country of origin. Some shows will accept competition entries only for the breeds and varieties officially recognized by the APA, but most shows will accept entries of rare breeds even if they are not in the *American Standard of Perfection*. Even those shows that do not accept nonstandard breeds for competitive classes normally will allow them to be entered for display.

Support Heritage Breed Organizations

There are a number of organizations that are dedicated to helping preserve rare breeds of livestock. They publish newsletters, sponsor seminars, and do a variety of promotional work. Supporting them with your membership can increase your knowledge and ability to work effectively for the good of heritage breeds. Appendix H provides contact information.

Help Standardize a Rare Breed

One step in helping increase the chances of survival of a breed is getting it recognized by the American Poultry Association. Once a breed is officially standardized, there is more incentive for specialty breeders and exhibitors to raise and show it, which in turn increases the breed's exposure to the public. Currently, the procedure for getting a breed standardized by the APA includes the following:

1. At least five people must raise the breed for a minimum of 5 years each.
2. A standard description of size, shape, and color is prepared for the breed.
3. At the end of the 5 years, at least five exhibitors must show a minimum of 50 specimens at a qualifying meet.
4. The judge of the qualifying meet must recommend to the APA board of directors that the breed be admitted due to sufficient uniformity as well as conformity to the written standard.
5. The APA board of directors votes on admittance.

While the task of getting a breed officially recognized sounds a bit arduous, it is feasible when a group of breeders pitches in to help.

The primary purpose of conserving rare breeds is for the benefit of future generations. Our niece Maggie, shown here at three years old, delights in watching over these day-old Saxony ducklings.

9

Hybrid Ducks

SOMETIMES FOR SPECIAL PURPOSES HYBRID DUCKS are produced by mating drakes and ducks of different breeds. Possible advantages of this scheme include the ability to combine characteristics that are not available in any one breed; the positive effects of heterosis (often referred to as hybrid vigor); and the use of sex-linked matings that produce offspring that can be sexed by color at hatching.

Heterosis

Heterosis refers to the difference in growth and production traits between crossbred offspring and the average of their parents. An example is the mating of a drake from a breed or strain that lays 300 eggs a year to a duck of a breed that normally lays 100 eggs a year. It would be logical to expect their offspring to produce 200 eggs per year. If, in fact, their female progeny produced 285 per year, the 85 "bonus" eggs could be credited to the positive effects of heterosis.

A side note: Heterosis occasionally has a negative effect on crossbred offspring, a fact that is sometimes overlooked. For example, when two breeds are crossed, their offspring often will have slightly higher body fat than the average of their parents. If your goal is to produce meat with a minimum of fat, this would be an example of a negative effect of heterosis.

Hybrids

All breeds of domestic ducks can successfully cross if there is not too great a difference in the lengths and sizes of their bodies. Generally, the breeds in the

Light-, Medium-, and Heavyweight classes can mate naturally with each other (both within and across these weight classes).

However, the greater the difference in body size between the breeds being crossed, the lower fertility usually will be. It is difficult for most bantam breeds to cross with ducks in the other three weight classes. Artificial insemination is sometimes employed to improve fertility or to make crosses between breeds that would be difficult or impossible for the birds to accomplish naturally.

Disadvantage of Hybrids

A major disadvantage of hybrids is that they often do not make suitable breeders. Some hybrids are sterile. When fertile hybrids are mated together, their progeny display great variability in their physical and production traits and a portion of heterosis is lost. Therefore, to produce hybrids, new purebred breeding stock is required for each generation.

Two-Way-Cross Hybrids

The most common method for producing hybrid ducks is to use a two-way cross. Drakes from breed (A) are mated to ducks from breed (B) to produce hybrid offspring (AB). Or, the reciprocal cross can be made in which drakes from breed (B) are mated to ducks from breed (A) to produce hybrid offspring (BA).

If you decide to produce hybrid ducks, it is important to understand that cross (AB) can look and perform differently from cross (BA). For example, when Khaki Campbell drakes from a high-producing strain are crossed with Black Cayuga ducks, their sons will be black, whereas the daughters will be chocolate and capable of laying approximately 300 eggs per year. In the reciprocal cross of Cayuga drakes × Khaki Campbell ducks, all the offspring are black and the daughters are capable of laying about 250 eggs per year.

Three-Way-Cross Hybrids

To combine the traits of three breeds or strains, drakes from breed (C) are mated with hybrid ducks (AB) to give progeny (CAB). Or the reciprocal cross — (AB) drakes mated to ducks from breed (C) to give progeny (ABC) — can be used.

These three-way-cross white hybrids were produced by mating White Campbell drakes to Pekin ducks. The F₁ offspring were then mated to White Runner ducks. The resulting offspring are quick growing, well muscled, and significantly leaner than Pekins, and the daughters are durable layers of white- or blue-shelled eggs.

Four-Way-Cross Hybrids

To combine the traits from four different breeds or strains, you first need to cross males (A) with females (B) to produce progeny (AB), and males (C) with females (D) to produce progeny (CD). In the second generation the (AB) birds are mated to (CD) birds to produce (ABCD) offspring. At each stage a reciprocal crossing can be made.

Sex-Linked Hybrids

There are situations when it is useful if ducklings can be sexed simply by looking at the color of their down, a phenomenon commonly referred to as sex-linked coloring. Sex-linked matings are made by mating sex-linked brown drakes (dark brown $Brsl^{br}Brsl^{br}$ or light brown $Brsl^{bu}Brsl^{bu}$) with non-brown females ($Brsl^+Brsl^+$). Interestingly, the mating of sex-linked light brown drakes with sex-linked dark brown females also gives sex-linked coloring in the resulting offspring. In all sex-linked matings, the sons will have the down color (the down pattern is irrelevant) of their dam, and the daughters will have the down color of their sire. The reciprocal crosses are not sex-linked.

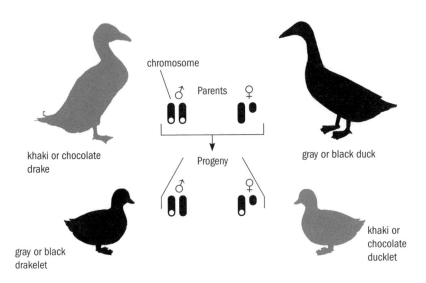

A brown drake crossed with a Mallard-colored or black duck is a sex-linked mating that produces gray or black sons and brown daughters.

Brown drakes include Khaki Campbell, Penciled Runner, Gold Harlequin, or Chocolate males of any breed (Fawn & White and Buff drakes can also be used, but they carry *blue*, which makes it more difficult to distinguish the sexes at hatching time, and therefore they are not generally recommended).

Females that can be used include Dark Campbell, Silver Harlequin, Silver Appleyard, and Blacks or Grays of any breed (technically, any blue-colored female can be used as well, but this is also not generally recommended). Most of the Pekins that I have tested do not have *brown* "hidden underneath" their white plumage and therefore can be used on the female side of a sex-linked cross.

At a poultry research station in the Caribbean during the 1970s, my students studied many sex-linked crosses as we searched for combinations that might produce useful birds for subsistence farmers in developing countries. We found that any cross using Khaki Campbell drakes out of high-producing strains produced marvelous laying ducks, no matter what breed their mates were (excluding Muscovy).

Khaki Campbell × Black Runner produced 3.5- to 4-pound (1.5 to 1.75 kg) hybrids that laid 290 to 320 eggs per hen per year and had unsurpassed foraging ability. Khaki Campbell × Cayuga produced 5.5-pound (2.5 kg) birds that laid 270 to 300 eggs per hen per year and were relatively calm and good foragers. Khaki Campbell × Pekin (a large commercial strain) produced progeny that weighed 6 to 7 pounds (2.75 to 3.2 kg) and laid 290 to 320 eggs per hen per year; these were the calmest of the crosses tried.

Harlequins, Runners, Magpies, and Anconas are also excellent sires of productive layers when crossed with other breeds of Mallard descent. Harlequin-sired hybrids lay about as well as Campbell hybrids, and their eggs are slightly larger. In general, hybrid daughters sired by Magpies lay the largest eggs of the hybrids discussed here.

Briefly, the reason these matings produce color-coded offspring is due to the position of the recessive *Brown* locus on the sex chromosomes. Male ducks are the homogametic sex because they carry a matched pair of sex chromosomes designated as ZZ. Female ducks are called the heterogametic sex since their two sex chromosomes, designated ZW, are of different lengths (the W chromosome is shorter). Note: In mammals, the male is heterogametic (XY) and the female is homogametic (XX). The *Brown* locus is located on that portion of the Z chromosome that is missing in the W chromosome. Therefore, a female duck can only inherit the *brown* or *buff* allele from her sire.

Muscovy Hybrids

When Muscovies are crossed with ducks of Mallard descent, the offspring are usually sterile "mules." This cross produces a wonderful meat bird that combines the fast growth rate of Mallard derivatives with the heavy muscling and leaner carcass of the Muscovy. These crosses can be beautiful and exotically colored.

Commercially, Muscovy drakes are crossed on Pekin, Rouen, or other large ducks. These eggs hatch in 4 weeks, and both males and females are a similar size at butchering age. The hybrid females from this cross (Muscovy drake × Mallard-derivative duck) typically do not lay, and the males are usually sterile.

The reciprocal cross of Mallard-derivative drake × Muscovy female produces offspring that take approximately 5 weeks to hatch and have considerable size difference between the sexes. The resulting hybrid females usually lay small eggs that do not hatch. Males from this cross are sometimes fertile.

For the best fertility, use Muscovy drakes that have not been with Muscovy females since an early age. Typically, one Muscovy drake for every five to eight ducks is recommended. In general, fertility for this cross runs at 40 to 60 percent when eggs are candled on the seventh day of incubation. Artificial insemination is commonly used by commercial producers of this cross.

10

Understanding Duck Colors

WITH A WORKING KNOWLEDGE OF plumage colors and patterns, you will have a valuable tool to help distinguish between the different pure breeds and to identify birds that have been either purposefully or mistakenly crossbred. Along with helping to indicate breed purity, colors have aesthetic appeal and important practical consequences, including camouflage from predators, appearance of dressed carcass, temperature control (white or light-colored ducks reflect more of the sun's rays), feather durability, and resistance to soiling.

The importance of color in heritage breed preservation should not be underestimated. Some people argue that only production qualities are important and color is superfluous. Throughout history there have been poultry breeders who ignored color and focused their breeding programs solely on "practical" qualities. Over time their lines of birds have been lost due to the lack of easy-to-identify plumage "markers."

Keep in mind that the genes responsible for color can also exert influence on "practical" qualities, such as disease resistance, growth rate, and fertility. For example, over the years I've had people report that ducks of a specific color thrive better in their unique microenvironment. While it can be difficult to pinpoint the gene(s) responsible, there is strong anecdotal evidence that there is a genetic component associated with that color that makes them better adapted to a given environment. For this reason my usual advice is that people try ducks of a variety of breeds and colors to see which perform the best in their specific situation. Case in point: When we sent Khaki Campbell and Magpie ducklings to Guatemala to be raised by native Kekchi subsistence farmers, I would never — in spite of working with agricultural development workers around the world for years — have thought that the Magpies would outperform the Campbells the way they did in that particular tropical ecosystem.

Basic Genetics

Having a rudimentary understanding of genetics is essential to understanding duck colors and how to improve them through breeding. The following is a brief review of the genetic principles required to understand basic color inheritance in ducks. (For detailed information refer to a textbook on Mendelian genetics. Better still, audit or take an appropriate genetics course.) Getting a grasp on the inheritance of colors takes time, study, and frequent review — unfortunately, many people give up before allowing themselves a chance to absorb the information.

Chromosomes

At a basic level genetics is the study of why an egg laid by a duck produces a duckling rather than, say, a duckbilled platypus. "Like begets like" because of the genetic blueprint carried on the threadlike chromosomes found in the nucleus of each cell.

Ducks have 80 chromosomes arranged into 40 pairs. A duckling inherits one chromosome of each pair from its dam, the other from its sire. There are 39 pairs of autosomal chromosomes in which the chromosomes of a given pair are matched for size. The 40th pair are the sex chromosomes. As mentioned in chapter 9, the male is the homogametic sex because it carries a matched pair of sex chromosomes designated as ZZ. The female bird is the heterogametic sex because with her two sex chromosomes, labeled ZW, the W is shorter.

Genes and Loci

Located along each chromosome are genes, which can be thought of as the codes or switches that control the characteristics of a duck. Some characteristics are controlled primarily by a single pair of genes (one gene located on each of a matched autosomal pair of chromosomes), whereas other characteristics are controlled by multiple pairs of genes. The specific site where a gene is found on a chromosome is called the locus (plural is loci).

Wild Type and Mutations

For any given trait the original form found in the wild ancestors of domestic ducks is known as the wild type. Understanding the concept of wild type is critical to understanding the inheritance of duck colors. The duck with her brood of ducklings pictured on the cover of this book is an excellent example of female and baby wild-type color in Mallard-derivative ducks (to see the wild-type nuptial and eclipse plumage patterns of drakes, see photos of the Mallard on page 117).

To avoid confusion keep in mind that Gray is the official name in North America for any duck that is colored exactly like a wild Mallard. The name was chosen by the American Poultry Association and the American Bantam Association because the predominant body color of wild Mallard drakes in nuptial plumage is gray. For example, when you see the name Gray Runner, think Mallard colored.

Occasionally, a gene (or chromosome) changes, which is known as a mutation. All domestic breeds of ducks, except those colored exactly like wild Mallards or wild Muscovies, carry at least one mutation gene for color. Mutated color genes can be added together in various combinations to make all the non-wild-type colors and patterns seen in domestic ducks. Another important concept to understand is that all non-wild-type colored domestic ducks carry both *wild type* and mutated genes.

Genetic mutations can affect size, color, plumage pattern, shape, disease resistance, skeletal formation, and more. A mutation can be beneficial, neutral, or harmful. There are some mutations whose impact is harmful or even lethal when the bird is homozygous for the mutation at that locus; whereas, its effect is neutral or even beneficial if the bird is heterozygous for that mutation.

Alleles

Alternative forms of a gene are called alleles. If a gene has one recognized mutation form, then it is said to have two alleles (the *wild type* plus the mutation form). If a gene has two recognized mutation forms, then it is said to have three alleles (the *wild type* plus the two mutation forms), and so on.

Homozygote and Heterozygote

If a duck carries a matched pair of genes at a particular locus, it is said to be homozygous for that locus. On the other hand, if it carries different alleles at a locus, it is called heterozygous. Breeders and hobbyists commonly use the terms "split" or "carrier" for birds that are heterozygous for color. For example, when attempting to improve the type of Snowy Calls by crossing onto Gray Calls with outstanding conformation, the first-generation (F₁) offspring would be phenotypically Grays, but would be described genotypically as heterozygous (or colloquially "Snowy carriers" or "split to Snowy").

Ducks are said to "breed true" when all the offspring resemble the parents from generation to generation. In order to breed true, both parents must be homozygous and have identical genotypes for the traits under consideration (if a spontaneous mutation occurs, one or more of the offspring may not resemble the parents).

Phenotype and Genotype

Phenotype refers to the outward appearance of an individual, whereas genotype is its actual genetic makeup. The phenotype of a duck's plumage is not always an accurate barometer of its genotype. For example, when you cross a Snowy Call with a Gray Call, the first-generation offspring normally look like Grays, often giving no visual clue that they are heterozygous for the *harlequin* allele (Snowys are homozygous for *harlequin*). So the phenotype of these crossbreds is Gray but their genotype includes one *wild type* allele and one *harlequin* allele. Mate these heterozygotes together and, typically, they will produce offspring in a phenotypic ratio of three Grays to one Snowy.

Dominant and Recessive Genes

Genes normally occur in pairs (one located on each matched chromosome). If a duck carries one *wild type* allele and one mutation allele of a gene at a given locus, and the color of the bird is not changed, then the *wild type* is said to be dominant and the mutation allele recessive. For example, when a homozygous Gray Mallard is crossed with a White Mallard, the offspring will be colored and will show no indication that they carry one *white* allele. Gray plumage is said to be completely dominant to white plumage. Despite being wild Mallard colored, these heterozygous ducks are *white* "carriers" and when mated together typically will produce offspring in a ratio of three Grays to one White.

The relationship of two alleles of a gene is not always so clear-cut. If a duck carries one *wild type* allele and one mutation allele at a locus and both alleles affect the appearance of the bird, then the relationship between these two alleles is incomplete dominance. *Blue* is an example of an incompletely dominant mutation. A Gray Call is homozygous for *wild type* at the *Blue* locus.

A Blue Fawn Call is heterozygous for the *blue* allele at the *Blue* locus. A Pastel Call is homozygous for the *blue* allele at the *Blue* locus. Notice that the phenotype is different in each of these three genotypes at the *Blue* locus.

Recessive genes can cause characteristics to "disappear from view" only to reappear in a later generation. For this reason experimental breeding programs should be continued for a minimum of two generations (I prefer three to five) to uncover all the genotypes present.

Sex Linkage

The loci of sex-linked traits are on the sex chromosomes. In birds, females are the heterogametic sex. If the locus for a gene is on the "missing" portion of the W chromosome, a female duck can only inherit it from her father — making it sex linked (see illustration on page 108.) Sex-linked matings can be used to produce ducklings that will be color coded for gender at hatching.

Genetic Ratios

When color genetics are discussed, various ratios are mentioned. For example, when a Blue drake is mated to a Blue duck, their offspring will hatch in a ratio of one Black to two Blue to one Silver or Silver Splash. These ratios are usually accurate when hatching hundreds or thousands of offspring. However, when hatching small batches of ducklings, do not be surprised if these ratios are off by a fair amount.

Genetic Nomenclature

At this writing there is no standardized nomenclature for poultry genetics. This situation can impede clear communication of current understandings of duck genetics. To help address this issue, a consistent format is used throughout this book that follows the guidelines set forth by the Committee on Genetic Nomenclature of Sheep and Goats, which has been expanded to include other domestic animals.

1. Names of loci begin with a capital letter; names of alleles begin with a lower case letter, except in unavoidable situations (such as the beginning of a sentence). Both loci and allele names are italicized. Example: The *restricted* allele is at the *Pattern* locus.

2. The symbols for loci are abbreviations of the names and are italicized and always have an initial capital letter. Example: The locus symbol for *Pattern* is *Pat*.

3. Allele symbols are added to loci symbols as superscripts to distinguish them from the locus symbol. The symbol $^+$ designates the probable *wild type* allele at a locus. Example: The *wild type* allele at the *Pattern* locus is written as *Pat$^+$*.

4. The symbols for alleles other than *wild type* are standardized so that dominant alleles have an initial capital letter; recessive alleles have only lower case letters. Both dominant and recessive alleles have an initial lower case letter when spelled out. Example: At the *Pattern* locus the *restricted* allele, *Pat^R*, is dominant to the *wild type* allele, *Pat$^+$*, but the *dusky* allele, *Pat^d*, is recessive to both *wild type* and *restricted*.

5. Another convention is the abbreviation of genotypes by using dashes to fill in behind dominant alleles when the second of a pair is unknown or unimportant. In purebred ducks this convention is seldom used because most color mutations are recessive or incompletely dominant, and therefore, the second allele is normally known and/or important for producing true color for the breed or variety described. However, for crossbred ducks of unknown ancestry, this convention is useful. Example: The genotype for Mallard-colored ducklings out of a mixed flock containing both white and colored ducks is could be written as *Pat$^+$-, C$^+$-*.

Additional adaptations that I have made in this book to clarify or simplify the nomenclature used:

6. The same locus and allele names and symbols are used in both Muscovies and Mallard-derived breeds when breeding tests indicate they are homologous. Example: In scientific literature *Nero* (*N*) is the name and symbol commonly used for Blue color in Muscovies. In this book the locus name *Blue* (*Bl*) will be used for both Muscovy and Mallard-derived ducks.

7. Unusual locus and allele names have been changed to ones that are more descriptive and intuitive. For example, in Muscovies, *Duclair Pied* has been changed to *Magpie* and *Canizie* to *White Head*.

8. In tables giving plumage color and pattern genes, the alleles are listed in order from dominant to recessive.

9. If a gene is known to be located on the sex chromosome, it will always be referred to as sex-linked. Otherwise, genes are assumed to be autosomal.

Mallard Derivative Colors

One of my favorite activities in raising ducks over the years has been observing and studying the marvelous array of plumage patterns and colors they exhibit. Because of this passionate interest, I have raised and studied ducks possessing each of the plumage patterns and colors covered in this section. Included are all the varieties that the American Poultry Association and the American Bantam Association recognize, plus most nonstandard varieties raised in North America. Two extremely rare colors that I have not yet been able to study are the Yellow Belly Call and a Mallard variety that goes unfortunately by the name "Silver"—but has no relationship to "Snowy" which is called "Silver" in Great Britain; neither is it related to the Silver discussed in this book. (For detailed descriptions of the standard varieties, see the latest editions of APA and ABA breed standards.)

The plumage colors and patterns discussed here are the result of *wild type* alleles, plus 13 mutations, as outlined in the table on page 118. Non-wild-type varieties are caused by anywhere from a single mutation to a combination of as many as five mutations.

In addition to the genes listed in the table, keep in mind that there are "modifying genes" that fine-tune plumage colors and patterns. These secondary genes are difficult to study and in general are poorly understood. However, the effects of these modifying genes are of special interest to breeders who want to improve their purebred stock.

To help understand how duck colors are "built," the varieties can be divided into six broad categories based on their genetic foundation:

1. Mallard Pattern: Blue Fawn, Butterscotch, Gray, Pastel, Saxony, Spotted, and Trout
2. Restricted Pattern: Silver Appleyard
3. Dusky Pattern: Aleutian, Buff, Cinnamon, Dark, Dusky, Harlequin, Khaki, Overberg, and Snowy
4. Extended Black: Black, Blue, Chocolate, Lavender, Lilac, and Silver
5. Self White
6. Pattern White: Ancona, Dominant Bib, Fawn & White, Magpie, Penciled, and Recessive Bib

Keep in mind that the genotypes given in the chart Foundation Plumage Pattern and Color Genotypes in Mallard-Derivative Ducks (on page 119) are for ducks that meet the breed standard for plumage pattern and colors. Some specimens that appear to have a given color may in fact have a different genotype that mimics, to varying degrees, a standard color.

THE PLUMAGES OF DUCKS

Ducks of Mallard descent have four distinct plumages during their lives: (1) ducklings are covered with furlike down; (2) at eight weeks of age, most of the down has been replaced with juvenile feathering, and both genders resemble adult females; (3) by sixteen to twenty weeks of age, young ducks sport their dimorphic nuptial (breeding) plumage, which is the plumage described in breed standards; and (4) at the end of each breeding season, both sexes molt into a juvenile-like eclipse plumage that is retained for 3 to 4 months.

At seventeen weeks of age, this Mallard drake has begun to replace his juvenile plumage with the bright colors of adulthood. To provide camouflage as long as possible, young males acquire their iridescent heads and white neck collars last.

A standard-bred Mallard drake in full nuptial plumage in early spring.

Here is the same drake as shown above but 4 months later in full eclipse plumage. By fall, he will again sport flashy nuptial plumage.

PLUMAGE PATTERN AND COLOR ALLELES
in Mallard-Derivative Ducks

Name	Symbol	Relationship to "Wild"	Main Visual Effects on Adults
Restricted Pattern	Pat^R	Dominant	Whitens surface of wing fronts; slightly lightens overall plumage
Wild Pattern	Pat^+	Wild type	Wild Mallard pattern
Dusky Pattern	Pat^d	Recessive	Eliminates facial stripes, obscures wing speculum; in drakes eliminates neck ring and claret breast
Sooty	So^{So}	Dominant	Darkens plumage of ducks with either *dusky pattern* or *wild pattern*
Wild	So^+	Wild type	Permits normal coloration
Wild	Li^+	Wild type	Wild Mallard coloration
Light	Li^{li}	Recessive	Moderately lightens plumage; in drakes extends claret onto shoulders and sides; sometimes enlarges neck ring
Harlequin	Li^h	Recessive	Further lightens plumage; enlarges wing speculum; in drakes extends claret onto shoulders and sides, darkens main tail feathers, neck ring often encircles neck
Wild, colored	C^+	Wild type	Permits pigment production in feathers
White	C^w	Recessive	Prevents pigment production in feathers
Extended Black	E^E	Incompletely dominant	Extends black throughout the plumage
Wild	E^+	Wild type	Permits normal plumage color expression
Blue	Bl^{Bl}	Incompletely dominant	Dilutes black pigment to blue gray in single dose; silver or silver-splashed white in double dose
Wild	Bl^+	Wild type	Permits normal plumage color expression
Wild	$Brsl^+$	Wild type	Permits normal plumage color expression
Brown, sex-linked	$Brsl^{br}$	Sex-linked recessive	Dilutes black pigment to dark brown
Buff, sex-linked	$Brsl^{bu}$	Sex-linked recessive	Dilutes black pigment to medium brown
Runner Pattern	Rn^{Rn}	Incompletely dominant	Prevents pigmentation of neck and parts of head, wings, and underbody
Wild	Rn^+	Wild type	Permits normal pigmentation throughout plumage
Dominant Bib	BiD^{Bib}	Incompletely dominant	Prevents pigmentation on front of chest/neck in ducks with *extended black*
Wild	BiD^+	Wild type	Permits normal pigmentation throughout plumage
Wild	BiR^+	Wild type	Permits normal pigmentation throughout plumage
Recessive Bib	BiR^{bib}	Recessive	Prevents pigmentation on front of chest in ducks with either *dusky pattern* or *wild pattern*

FOUNDATION PLUMAGE PATTERN AND COLOR GENOTYPES
in Mallard-Derivative Ducks

Plumage Pattern	Color Genotypes
Gray (wild type, Mallard)	Pat^+Pat^+, So^+So^+, Li^+Li^+, C^+C^+, E^+E^+, Bl^+Bl^+, $Brsl^+Brsl^+$, Rn^+Rn^+, BiD^+BiD^+, BiR^+BiR^+
Pastel	$Bl^{Bl}Bl^{Bl}$
Trout, Claire	$Li^{li}Li^{li}$
Saxony, Butterscotch	$Li^{li}Li^{li}$, $Bl^{Bl}Bl^{Bl}$
Greenhead Australian Spotted	$Li^{li}Li^h$ (possibly others from wild species)
Silver Appleyard	Pat^RPat^R, $Li^{li}Li^{li}$
Dusky	Pat^dPat^d
Aleutian, Dark Silver	Pat^dPat^d, $Li^{li}Li^{li}$
Khaki	Pat^dPat^d, $So^{So}So^{So}$, $Brsl^{br}Brsl^{br}$
Light Khaki (rare)	Pat^dPat^d, $So^{So}So^{So}$, $Brsl^{bu}Brsl^{bu}$
Dark (Campbell)	Pat^dPat^d, $So^{So}So^{So}$
Solid Fawn, Faery Fawn	Pat^dPat^d, $So^{So}So^{So}$, $Li^{li}Li^{li}$, $Brsl^{br}Brsl^{br}$
Gold Harlequin	Pat^dPat^d, $So^{So}So^{So}$, Li^hLi^h, $Brsl^{br}Brsl^{br}$
Silver Harlequin	Pat^dPat^d, $So^{So}So^{So}$, Li^hLi^h
Snowy	Pat^dPat^d, Li^hLi^h
Overberg	Pat^dPat^d, Li^hLi^h, $Bl^{Bl}Bl^{Bl}$
Buff	Pat^dPat^d, $So^{So}So^{So}$, $Bl^{Bl}Bl^{Bl}$, $Brsl^{bu}Brsl^{bu}$
Black	Pat^dPat^d, E^EE^E
Blue	Pat^dPat^d, E^EE^E, $Bl^{Bl}Bl^+$
Silver	Pat^dPat^d, E^EE^E, $Bl^{Bl}Bl^{Bl}$
Dark Chocolate	Pat^dPat^d, E^EE^E, $Brsl^{br}Brsl^{br}$
Medium Chocolate	Pat^dPat^d, E^EE^E, $Brsl^{bu}Brsl^{bu}$
Lavender	Pat^dPat^d, E^EE^E, $Bl^{Bl}Bl^+$, $Brsl^{br}Brsl^{br}$
Lilac	Pat^dPat^d, E^EE^E, $Bl^{Bl}Bl^{Bl}$, $Brsl^{br}Brsl^{br}$
White	C^wC^w (plus any "hidden" colors and patterns)
White Bibbed Black	Pat^dPat^d, E^EE^E, $BiD^{Bib}BiD^{Bib}$
White Bibbed Gray (Mallard)	$BiR^{bib}BiR^{bib}$
White Bibbed Dusky	Pat^dPat^d, $BiR^{bib}BiR^{bib}$
Penciled	Pat^dPat^d, $So^{So}So^{So}$, $Brsl^{br}Brsl^{br}$, $Rn^{Rn}Rn^{Rn}$
Fawn & White	Pat^dPat^d, $So^{So}So^{So}$, $Bl^{Bl}Bl^{Bl}$, $Brsl^{br}Brsl^{br}$, $Rn^{Rn}Rn^{Rn}$
Pied	$Rn^{Rn}Rn^{Rn}$
Black Magpie	Pat^dPat^d, E^EE^E, $BiD^{Bib}BiD^{Bib}$, $Rn^{Rn}Rn^{Rn}$
Black Ancona	E^EE^+, $BiD^{Bib}BiD^{Bib}$, $Rn^{Rn}Rn^{Rn}$

Bill, Skin, and Leg Color

Surprisingly little has been written about the bill and skin color of ducks. The colors of these areas are important breed characteristics and, in the case of legs and bill, are part of their standard description. Furthermore, skin color can help determine the feasibility of a breed or variety for specific markets and can be an important clue to special dietary requirements.

Skin covers the feet, legs, body, and bill of ducks, and its color is genetically controlled. However, the intensity of color is strongly influenced — sometimes dramatically so — by diet, environment, natural hormone fluctuations during the different seasons of the year, and the health of the individual.

The wild-type skin color is yellow. The legs and feet of drakes are orange to reddish orange, while those of females are brownish orange. Ducks have the richest-colored bills, skin, and legs when they are in glowing health, are consuming a well-balanced diet that is rich in yellow xanthophylls, are exposed to adequate sunlight, and prior to heavy egg production. As with all birds, yellow pigment is pulled from the laying duck's skin and deposited in the yolk, leaving her bill and skin paler the longer she lays.

Understanding bill and skin color in white ducks is straightforward if it is kept in mind that the two are inextricably linked. Ducks with pink bills have pink skin; ducks with yellow bills have yellow skin. Ducks with pink bills and skin, such as Aylesburies and White Hook Bills, have a dominant allele that I call *pink* at the *Skin Color* locus (Skn^P) that greatly reduces the amount of yellow dietary pigment that is deposited in the skin and body fat. (Interestingly, pink-billed ducks still have orange legs and feet.) In some localities people will pay a premium for pink-skinned meat ducks, whereas in other areas they are not preferred. So if you plan on marketing meat ducks, it behooves you to know the preference of your clientele before choosing your breed. Pink-skinned ducks typically need higher levels of vitamins D_3 and A than do yellow-skinned varieties for overall health and to prevent leg weakness and poor fertility.

White ducks that have yellow bills and skin are homozygous for the *wild type* allele at the *Skin Color* locus (Skn^+Skn^+) that allows yellow xanthophylls to be deposited normally. When white ducks with yellow skin eat sufficient quantities of xanthophylls, even their newly grown feathers are pigmented creamy yellow.

The genetics of bill color in ducks with colored plumage can seem complex, in part because bill color/markings of males and females within the same variety often are dramatically different. Taking it one step at a time will go a long way toward full comprehension. To understand bill color in colored ducks, it is important to remember that there are several layers of pigmentation in the bill. One layer has either yellow-orange pigment (Skn^+Skn^+) or no pigment (Skn^P-). A second layer has blue pigment.

Wild-type colored drakes have yellowish green to green bills due to the combination of yellow and blue layers. Colored drakes that have the *pink* allele (Skn^P) — such as Dusky Hook Bills — have blue bills due to the lack of yellow pigment. (Blue-billed drakes that are eating large quantities of yellow pigment will typically have at least a bit of green shading in their mostly blue bills due to a bit of yellow pigment "leakage.")

In colored female ducks the original bill color (Skn^+Skn^+) is orange to brownish orange with a nearly black saddle that can cover anywhere from 5 to 95 percent of the bill surface. The orange portions of the wild-type female's bill are the brightest orange in the fall and early winter months, when ducks are in their full nuptial colors and pairing off for the next breeding season. As the laying season approaches and progresses, the dark portion of the female's bill typically increases and sometimes ends up covering the entire upper mandible. If a colored female duck has the *pink* allele (Skn^P), the orange portions of her bill will be brownish pink with the normal dark saddle markings.

Now things get even more interesting! In some varieties of domestic ducks (e.g., Khaki Campbell and Cayuga), the female's bill color is a solid dark color. I call the locus that determines whether a female's bill will have the saddle pattern or a uniform dark color *Bill Pattern (Bil)*. The s*olid bill* allele, Bil^{Sd}, is dominant to the *wild type* (saddle pattern) allele, Bil^+. Most people call the bill color of these solid-bill-pattern females "black"; however, upon close examination, it can be seen that these bills are actually either dark green or dark blue. Those females that have yellow skin will have dark green bills (Skn^+Skn^+, Bil^{Sd}-), while those with pink skin will have dark blue (Skn^P-, Bil^{Sd}-).

Keep in mind that in many varieties the *solid bill* allele does not visually affect the bill color of drakes. Therefore, it is not possible in many cases to determine the bill pattern genotype of a drake without doing breeding tests. However, in varieties that have *extended black* E^E, the *solid bill* allele does increase the amount of dark color in the male's bill.

Gray

The plumage pattern and colors of the wild Mallard are officially known as Gray in North America (interestingly, they are known as Brown in Great Britain). Except where noted, all other color varieties are thought to be mutations of the Gray. Genotypically, authentic Grays are homozygous *wild type* at every pattern and color loci. Breeds that have a Gray variety include Call, Mallard, Rouen, Runner, and Silkie. (For genotype see Foundation Plumage Pattern and Color Genotypes in Mallard-Derivative Ducks chart, page 119.)

Day-old description. Ducklings have the classic black and yellow camouflage pattern (see front cover photo). Greenish black runs from the base of the bill, over the top of the head, and down the back of the neck, then spreads out across the shoulders, back, and tail. The face is yellow with one or two dark facial stripes, and the back and wings are marked with yellow "sun spots." The front of the neck, chest, and underbody are yellow. The bill is black, often with yellow shading on the margins. The feet and legs are gray yellow with dark webs. When these ducklings hold still in grass, they become nearly invisible as their colors blend in with sunspots and shadows.

Adult description. The drake in nuptial plumage (see photo, page 117) has an iridescent green head, white neck collar that does not meet in the back, claret chest, steel gray underbody, black rump and under-tail coverts, white to ashy gray tail feathers, and reddish orange feet and legs. Depending on the time of year, the bill varies from yellow to olive green with a black bean tip.

As illustrated in the cover photo, the female has intricate feather patterns and lovely earth tones, which give her superb camouflage when nesting. Her base color can vary from golden brown to rich mahogany, with each feather distinctly penciled with dark brown. The legs and feet are brownish orange and the bill is orange, with a dark saddle across the middle and a black bean at the tip. Bill color typically darkens during the breeding season for enhanced camouflage. The bill of the female pictured on the front cover most likely was 60 to 80 percent orange several months prior to laying her clutch of eggs. Bright-colored bills are a natural part of nuptial coloring in wild-type ducks. Both sexes have brown eyes and iridescent wing speculums, which, depending on the lighting, can be brilliant blue to purple.

Juvenile drakes resemble females but can often be identified by their green bills and darker color, with less penciling on the top of the head and lower back. The bright colors of the nuptial plumage gradually begin appearing at ten to twelve weeks and are completed by sixteen to twenty weeks of age. Juvenile females have duller, less distinctly penciled plumage than adults. For several months each year (usually in late spring and summer), drakes exchange their bright hues for subdued, female-like eclipse plumage as they replace their flight feathers.

Common faults. For exhibition the following are undesirable in drakes: clear yellow bill, foreign color on the head, poorly defined white neck ring, light lacing on claret feathers of the chest (this "chain armor" typically is most noticeable on newly grown feathers and then darkens), muddy steel gray on body, white or gray mixed in with the black of the under-tail coverts, and black on the ridge of bill (most Calls have this fault).

In the duck—as much as possible—avoid indistinct facial stripes, poor penciling in body plumage and wing bows, faded color on the underbody and under-tail coverts (usually, even the best exhibition Call females are pale in under-tail covert color), solid white feathers on front of neck, and solid lead-colored bills (because of hormonal changes it is normal for the bill to darken prior to and during the laying season). In both sexes watch for poorly colored wing speculum with indistinct black and white border on top and bottom.

Breeding hints. In exhibition Rouen drakes it is preferred that there not be a vertical white border of feathers between the steel gray of the flank and the black of the under-tail coverts and rump. However, males that have this "cotton" as well as "chain armor" lacing on their chests normally produce daughters with the most distinct penciling. Also, drakes with the most distinct penciling in their juvenile plumage usually produce daughters with strong penciling.

Pastel and Blue Fawn

Pastel is a soft color that is known as Apricot in Great Britain. The plumage pattern and color genotype is *wild type*, except for two incompletely dominant *blue* genes: $Bl^{Bl}Bl^{Bl}$. Blue Fawns are heterozygous for *blue* — $Bl^{Bl}Bl^+$ — and in both day-olds and adults are darker and bluer than Pastels. Skin and bill color genotypes are *wild type*, Skn^+Skn^+, Bil^+Bil^+. Breeds that have Pastel and Blue Fawn varieties include Call, Crested, Mallard, Runner, and Rouen.

Day-old description. Pastel ducklings have the same pattern as Grays, but all black sections of the down are diluted to silver. The bill is grayish brown, and the legs and feet are grayish yellow to brownish yellow.

Adult description. Pastel drakes have the same pattern as Grays, but the head, neck, rump, and under-tail coverts are diluted to soft powder blue, and the body is pale silver gray with a slight buff or rose hue. Depending on the time of year, the bill varies from yellow to green.

Overall, the duck is a blend of silver, buff, and salmon, often with a rose hue. The top of the head, back of the neck, and eye stripes are silver or brownish gray, and the bill is orange with a dark brown saddle. In both sexes the wing speculum is bluish gray, bordered on top and bottom with white, and the legs and feet are orange to salmon.

Common faults. Pastels can have the same pattern faults as the Grays. Drakes often have foreign color in their heads and some ducks have nearly solid silver bodies without the desired buff, salmon, and rose overtones.

Breeding hints. If you need to improve type in Pastels, mate them to Grays with excellent conformation and rich brown plumage (excessively dark-colored Grays produce Pastel females with nearly solid silver bodies). Pastel × Gray produces all Blue Fawn offspring. For the second generation mate together the best-typed Blue Fawns with rich brown tones in their juvenile plumage, and they will produce offspring in the ratio of one Gray to two Blue Fawn to one Pastel. When mated together, these Pastels will breed true.

Day-old exhibition Rouens in three colors, from left to right: Pastel, Blue Fawn, and Gray.

Saxony, Butterscotch, and Trout

Saxony is a soft color named for the large German Saxony breed. The plumage pattern and color genotype is *wild type* except for two incompletely dominant *blue* and two recessive *light* genes: $Bl^{Bl}Bl^{Bl}$, $Li^{li}Li^{li}$. A common mistake is to assume that Saxony is a variant of the Buff color, but genetically they are distinct colors. Skin and bill color genotypes are *wild type*, Skn^+Skn^+, Bil^+Bil^+. Besides the German Saxony, this color is being bred in Call and Runner.

Butterscotch Call ducks present an interesting situation. Females possessing the desired color have the Saxony genotype, but because the *light* allele is variable in its expression, breeders have been able to select for more creamy white in the underbody, flanks, and facial plumage in the Butterscotch. Drakes that are winning in show competition have variable genotypes: one is Saxony (these tend to show more foreign color in the head but produce daughters with the desired distinctive facial stripes), while others have one or two *harlequin* alleles (these often have cleaner head color but produce daughters lacking well-defined facial stripes).

The Trout color genotype, $Li^{li}Li^{li}$, is the same as for Saxony, minus the two *blue* genes.

Day-old description. Saxony and Butterscotch ducklings are similar to Pastels but overall are lighter and a bit yellower in color.

Adult description. At first glance the Saxony and Pastel look similar. Upon closer examination you will find that there are a number of distinctions.

In Saxony drakes, the claret breast color is larger and bleeds back onto the shoulders and sides of the body and is frosted or laced with cream, especially on the lower portion. The oatmeal-colored body is paler than in the Pastel. Ideally, the Saxony drake's neck ring fully encircles the neck, although most drakes have a slight gap at the back. The head, neck, rump, and undertail coverts are diluted to a soft powder blue. The bill is yellow, with varying amounts of green shading, depending on the time of the year.

The plumage of the Saxony duck is a warm fawn to cinnamon buff with distinct creamy white facial stripes, throat, and neck front. The bill varies from yellow to brown, depending on the season of the year. Both sexes have silver-gray wing speculums bordered on top and bottom with white and orange feet and legs.

Common faults. Foreign color mixed in with the powder blue of the head and neck is especially a problem in older drakes. Brown-colored heads, indistinct white neck rings, and undersized claret bib on the chest in males are highly objectionable; in females, lack of creamy white throat, neck front, and facial stripes are unacceptable in shows.

Breeding hints. When choosing breeders, give priority to males with clean head color, distinct white neck rings, large claret bibs, and bills that are as yellow as possible. In females look for warm fawn buff color overall, offset by distinct creamy white on the throat, neck front, and face. Using some extra-dark drakes and rich cinnamon fawn females in the breeding pen can help restore faded overall coloring. In Butterscotch Call females a paler underbody color is desired.

Comments. There are two forms of pseudo-Saxony coloration that can wreak havoc if they are mistakenly used in the breeding pen of true Saxony. The genotype of Type I pseudo-Saxony is $Pat^{d}Pat^{d}$, $Li^{li}Li^{li}$, $Bl^{Bl}Bl^{Bl}$, $Brsl^{bu}Brsl^{bu}$ and is produced as a color variant by some strains of Buff Orpingtons. Phenotypically, these drakes often have poorly defined claret bibs, muddy white neck rings, and brown- or fawn-colored heads. Ducks have either solid fawn heads and necks or poorly defined facial stripes.

Type II pseudo-Saxony have a genotype of Pat^dPat^d, $Li^{li}Li^{li}$, $Bl^{Bl}Bl^{Bl}$ which is found in Runners and crossbred "pond" ducks from time to time. Drakes can closely mimic true Saxony but often have poorly defined claret bibs and muddy white neck rings, while ducks have solid-colored fawn heads and necks.

The reason for these pseudo-Saxony colorations is that when a pair of *light* genes is combined with *dusky* (Pat^dPat^d, $Li^{li}Li^{li}$), the claret breast and white neck ring of the drake are at least partially restored. These imitation Saxony are sometimes promoted under the label of "American Saxony."

Australian Spotted

The plumage pattern and colors of the Australian Spotted have proven to be a challenge to study — possibly because unnamed genes from the Northern Pintail and a wild Australian species may also be present (see chapter 4 for the history of the Australian Spotted). Australian Spotted offspring often segregate into three plumage patterns: spotted, snowy, and trout. The individuals with the most dramatically spotted plumage have a pair of the *wild pattern* alleles plus single alleles for *light* and *harlequin*, giving it a plumage pattern and color geneotype of Pat^+Pat^+, $Li^{li}Li^h$ for the Greenhead variety. Australian Spotted are being bred in Greenhead, Bluehead, and Silverhead varieties. Blueheads are heterozygous for *blue*, and Silverheads are homozygous for *blue*. Skin and bill color genotypes are *wild type*, Skn^+Skn^+, Bil^+Bil^+.

A historical note: In the 1970s several old-time breeders told me that Australian Spotted were used in the development of Snowy, Spotted, and Aleutian Calls.

Day-old description. Greenhead ducklings resemble wild Mallards, with a bit more yellow in their down. In Blueheads the black portions of the down are blue gray, whereas in the Silverheads the black sections are silver.

Adult description. Greenhead drakes resemble Mallards, except the head is sometimes marked with more or less brown and gray, the white collar fully encircles the neck, the claret of the breast extends back onto the shoulders and sides of the body, and the underbody is a pale, creamy silver that extends up onto the front of the chest. Depending on the time of year, the bill is yellowish green to green. Drakes in eclipse plumage display conspicuous black or brown spots on much of their body plumage.

Greenhead ducks are light to medium fawn (the underbody, lower back, front, face, neck, and a faint neck ring are paler), with most feathers having an elongated dark brown spot. On the head dark-brown stippling is most prominent across the crown and in a stripe through the eyes. The bill is orange and black. Both sexes have iridescent blue wing speculums and orange legs and feet with varying amounts of gray shading.

In Blueheads all black and dark brown portions of the plumage are diluted to blue gray, whereas in Silverheads all black and dark brown portions of the plumage are diluted to silver.

Common faults. Due to the genetic makeup of this color, it is normal for there to be a range in the appearance of Spotted colored ducks. Common faults include indistinct white neck ring and lack of claret on the shoulders and sides of the body of drakes and excessively dark underbodies and indistinct spotting in the body plumage of ducks.

Breeding hints. For breeding choose drakes with extensive claret on their shoulders and body sides when in nuptial plumage and bold spots in their juvenile and eclipse plumages, and ducks with distinctly spotted feathering. The subvarieties can be mated in any combination with the following results:

1. Greenhead × Greenhead produces all Greenhead offspring.
2. Greenhead × Bluehead produces half Greenheads and half Blueheads.
3. Greenhead × Silverhead produces all Blueheads.
4. Bluehead × Bluehead produces all three colors in a ratio of one Green to two Blue to one Silver.
5. Bluehead × Silverhead produces half Blueheads and half Silverheads.
6. Silverhead × Silverhead produces all Silverheads.

Silver Appleyard

The Silver Appleyard color is named after the breed of the same name. The plumage pattern and color genotype is *wild type* except for two dominant *restricted pattern* and two recessive *light* genes: $Pat^R Pat^R$, $Li^{li} Li^{li}$. This is the only officially recognized color that carries the *restricted* allele. The Silver Appleyard breed is the only one that I know of with this genotype. (There are Runners and Calls that are labeled Silver Appleyard colored; however, the ones that I am aware of have the *light* allele but not the *restricted* and therefore are not true Silver Appleyard color.) Skin and bill color genotypes are *wild type*, Skn^+Skn^+, Bil^+Bil^+.

Day-old description. Ducklings have smoky yellow down with a black "Mohawk" crown and black lower back and tail. Their bills are yellow with an irregular black stripe down the middle, and their feet are yellow with more or less gray shading.

Adult description. Drakes resemble Gray Mallards with the following differences: silver-white throat and face flecked with fawn, white neck collar that fully encircles the neck, claret chest color lightly frosted with white that extends onto shoulders and along sides of body, and pale silver and cream underbody. Appleyards have main tail feathers that are creamy white, shaded with pale gray. Bills are greenish yellow, and legs and feet are orange.

Females are light-Mallard colored, with white extending from the throat down the front and sides of the neck, onto the chest and underbody, and back to the abdomen. Running from the top of the bill across the crown of the head, down the back of the neck, and spreading across the shoulders and onto the sides of the body is an unbroken band of fawn feathers, center-marked with dark brown. The face is white with a faint eye stripe. The surface of the wing fronts is white. The plumage has a strong tendency to darken as the female ages. Bills are orange to brownish orange with a dark blotch on the ridge (they typically darken during the laying season), and the legs and feet are orange to brownish orange.

Day-old Silver Appleyards with typical dark tails and "Mohawk" head markings against a yellow background. The dark head and tail markings are not this distinct in all Appleyard ducklings.

Common fault. In drakes smoky black main tail feathers and solid black throats are serious faults, indicating that they are not genotypic Appleyards. In ducks excessive color on the throat, neck front, chest, and underbody; lack of white surface on wing fronts; fawn band down back of neck broken with white; and greenish black bills are faults to avoid for exhibition birds. Many females with the overall best markings have dull or small iridescent wing speculums and should not be severely penalized for breeding or showing.

Breeding hints. Drakes with the most silver white in their head plumage produce daughters with the best white-fronted color pattern. When possible, avoid breeding from drakes with solid-green heads and smoky black tails and ducks with solid-green or -black bills. When selecting breeding stock, keep in mind that birds with the whitest chests and underbodies will dress off the cleanest when butchered for meat.

Dusky and Sooty

A quick review of genotypes (see Plumage Pattern and Color Alleles in Mallard-Derivative Ducks, page 118) reveals that *dusky* is a foundation allele at the *Pattern* locus in many duck colors. Beginning with R. George Jaap (July 1934, Alleles of the mallard plumage pattern in ducks. *Genetics.* 19:310–322), researchers of duck colors have classified *dusky* as recessive to *wild type*. I found this puzzling, since my work of nearly four decades with *dusky* showed it to be incompletely dominant to *wild type* in F_1 heterozygous individuals.

Then in 2001 I began studying a new line of dusky-colored ducks that indeed proved to be recessive to *wild type*. Further work has shown that the old lines of Khaki Campbells, Buff Orpingtons, and certain colors of Runners (and some derivatives of these breeds) that I used for my early studies of duskies have an additional genetic component, which I call *sooty*. *Sooty* in conjunction with *dusky* makes *dusky* appear to be incompletely dominant to the *wild type* allele.

Day-old description. With practice you can identify the genotype of ducklings possessing different combinations of *dusky* and *sooty*. (1) Homozygous *dusky* (Pat^dPat^d) ducklings have uniformly off-black down and leg, foot, and bill color that is darker than those of *wild type*; (2) Ducklings that are heterozygous for *dusky* (Pat^+Pat^d) have the classic black and bright yellow camouflage pattern, and are indistinguishable from homozygous *wild type*; (3) Individuals that are homozygous for *dusky* and *sooty* (Pat^dPat^d, $So^{So}So^{So}$) have the solid dark down of the #1 ducklings above but usually with even darker leg, foot, and bill color; (4) Ducklings that are heterozygous for both *dusky* and *sooty*

(Pat^+Pat^d, $So^{So}So^+$) have modified *wild type* down color and pattern, with the light portions being dull grayish yellow rather than the brilliant yellow seen in *wild type* ducklings.

Adult description. Dusky drakes in nuptial plumage are distinctive, with their light to medium bluish gray bodies sandwiched between the iridescent, green-black head and neck and lower back and under-tail coverts. Females, juvenile drakes, and adult drakes in eclipse plumage have no facial stripes and are uniformly colored throughout their entire plumage. Both genders have dark gray wings with partially obscured wing speculums.

Drakes in nuptial plumage that carry *sooty* are several shades darker, have duller iridescent portions, and often have a brownish hue to their plumage. Females of all ages, juvenile drakes, and those in eclipse plumage are a shade or two darker than individuals that do not carry *sooty*. Foot and leg color in both genders and all ages is darkened.

Comments. Ducks that carry *dusky* have a strong inclination to produce at least some offspring with white on the front of the neck and/or under the throat at the junction of the bill and feathers. In purebred breeding programs

This Dutch Hook Bill drake, bred on our farm, is an excellent example of the *dusky* pattern (without *sooty*). Note: He happens to be in an upright startled pose.

it is often advisable to avoid breeding from those birds with white, since it increases from generation to generation.

Sooty is important in khaki-, buff-, and solid fawn-colored varieties because it enhances the uniformity of body plumage color and helps produce browner head, neck, speculum, lower back, and under-tail covert color. It also darkens leg and foot color.

Khaki and Dark

The warm earth tones of Khaki plumage give ducks of this variety excellent camouflage. The plumage pattern and color genotype include three non *wild type* alleles: *dusky, sooty,* and *sex-linked brown*: $Pat^d Pat^d$, $So^{So} So^{So}$, $Brsl^{br} Brsl^{br}$. (dark Campbells have the same genotype as Khakis, minus the *sex-linked brown* allele). *Sooty* reduces the amount of green iridescence in the head and neck of drakes and slightly darkens and increases the uniformity of plumage color throughout the body. A genetic quirk of the *dusky pattern* causes some offspring to have more or less white on the front of the neck and under the bill — even if breeding birds with these mismarkings are never used. Ducks of this variety do not have true khaki color until their plumage is faded from exposure to the elements. In North America the preferred skin and bill color genotypes are $Skn^+ Skn^+$, $Bil^{Sd} Bil^{Sd}$. In Great Britain the preferred is $Skn^P Skn^P$, $Bil^{Sd} Bil^{Sd}$. Breeds that have the Khaki variety include Call, Campbell, and Runner. Campbell is the only breed with the Dark variety.

Day-old description. Khakis have olive brown down with greenish or bluish brown bills and brown feet and legs. Darks are nearly black.

Adult description. In drakes the head and neck, lower back, wing speculum, and under-tail coverts are brownish black with a hint of green iridescence, and the body is an even shade of medium to tannish brown. Depending on the time of year, the bill is greenish yellow to green (greenish blue is preferred in Great Britain), and the legs and feet are orange. In ducks the body is brown (with varying amounts of penciling), the head and neck are a shade darker than the body, the bill is greenish black (bluish black in Great Britain), and the feet and legs are brown to orange brown (for exhibition, drakes with bronze heads and green bills are preferred, as are ducks with a minimum of penciling in their body plumage and wing speculums that match the body color).

Common faults. Ducklings or adults that have facial stripes are crossbreds and should not be used in a purebred breeding program or for exhibition. Also, avoid drakes with distinct white neck collars and claret on the chest.

Breeding hints. To produce the best percentage of even-colored birds for exhibition, identify potential breeding drakes at seven to twelve weeks of age that show the least penciling in their juvenile plumage. Because white increases from one generation to the next, whenever possible, avoid breeding from birds of either sex that have white on the neck or on the chin at the junction of the lower mandible and throat.

Harlequin

This high-contrast color pattern is named after the Welsh Harlequin breed. In North America there are two color varieties, Silver (the most common) and the original Gold (the only recognized variety in Great Britain). The plumage pattern and color genotype of Gold Harlequins is the same as Khaki, with the addition of two recessive *harlequin* alleles: $Pat^d Pat^d$, $So^{So} So^{So}$, $Li^h Li^h$, $Brsl^{br-} Brsl^{br}$. Silvers lack the *sex-linked brown* allele. Skin and bill color genotypes are $Skn^+ Skn^+$, $Bil^{Sd} Bil^{Sd}$.

Day-old description. Gold Harlequin ducklings are yellow, with more or less brown blush on the head and tail. Silver Harlequin ducklings are yellow with more or less gray blush on their head and tail. For the first couple of days after hatching, both phases can be sexed with about 75 percent accuracy by bill color: most drakelets have gray or dark green bills, while ducklets usually have yellow or tan bills with a dark tip (some Harlequins carry modifying genes that lighten the bills of drakelets and darken the bills of ducklets). Legs and feet are gray or green with yellow shading.

Adult description. The iridescent green–headed Silver drake resembles a Mallard that has been detailed by an artist, with the heavily frosted claret of his breast extending well back onto the shoulders and sides of his body. A white collar fully encircles his neck, getting wider toward the back. The distinctive tail color is a diagnostic marker of authentic Harlequin drakes: the main tail feathers are smoky black, each with a well-defined, narrow white border. Bill color is greenish yellow to green, depending on the time of year. Feet and legs are orange, shaded with brown.

Silver ducks are light to medium fawn, frosted with white, and each body feather is center-marked with dark brown. The head and neck are frosted white and shaded with light to medium fawn, similar to the body but with darker brown stippling especially noticeable on the crown of the head. The bill is greenish black, and the legs and feet are orange brown. In both sexes the wing speculum is bright blue to purple (depending on the lighting) and often extends out onto the tertial feathers.

Golds are similar to the Silvers but with slightly muted plumage colors. In Golds the head, neck, lower back, and under-tail coverts of the drake have a slight bronze hue; in ducks the dark center marks of the plumage are diluted to medium brown. In both sexes the wing speculums are greenish bronze rather than the bright blue of the Silvers.

In both color varieties it is common for drakes, especially those more than a year old, to have a bit of silver and fawn flecking in their head color, especially around the eyes and in the feathering covering the ears.

Common faults. In drakes obscure white neck collars; indistinct white frosting or lacing on the feathers of the chest, shoulders, upper back, and sides of the body; and light-colored main tail feathers are faults to be avoided. In ducks light-colored bills; head and neck color darker than the rest of the body (this is a major difference between the Snowy and Harlequin — in Snowys a darker head and neck color is highly desirable); body feathers lacking dark center marks; head, neck, and body totally lacking fawn blush; and poorly defined wing speculum are common faults.

Breeding hints. Diagnostic characteristics of Harlequins include black main tail feathers edged with white in drakes and no distinct facial stripe in ducks. Do not breed from drakes with light-colored tail feathers or ducks with facial stripes.

Snowy and Overberg

Snowy is a high-contrast pattern called Silver in Great Britain. Originally, the majority of Snowys were *wild type* except for two recessive *harlequin* alleles, Li^hLi^h. Today Snowys that are winning at shows also have the recessive *dusky pattern*, giving them the genotype Pat^dPat^d, Li^hLi^h. *Dusky* enriches the colors of both drakes and ducks and gives females the desired fawn hood. Skin and bill color genotypes are Skn^+Skn^+, Bil^+Bil^+. Breeds that have a Snowy variety include Call, Mallard, and Silkie.

While the original Miniature Silver Appleyards developed by Reginald Appleyard and being bred in North America today have the Snowy genotype, the variety name "Snowy" is not used.

The Overberg color genotype is the same as Snowy, with the addition of two incompletely dominant *blue* genes.

Day-old description. These yellow ducklings look like they've been airbrushed with coal dust. The gray blush is usually heaviest on the head, lower back, and tail. For the first couple of days, ducklings can be sexed with fair accuracy by bill color — most drakelets have gray or dark green bills, while the majority of ducklets have yellow or tan bills with a dark tip. Legs and feet are gray or green with yellow shading.

Adult description. Snowy plumage displays considerable variation in its expression among individual birds. Both sexes resemble their Mallard counterparts, except most of their body feathers are heavily edged with white. In drakes the white neck collar fully encircles the neck, the heavily frosted claret coloration extends onto the shoulders and sides of the body, and the main tail feathers are smoky black with a narrow white border. The bill is greenish yellow to green, depending on the time of year. The feet and legs are orange shaded with brown.

Ideally, ducks have a distinct fawn-colored hood (head and upper neck), which is in clear contrast to the rest of the body plumage. The body is heavily frosted with white, each feather center-marked with brown. Bills are brownish orange with a dark saddle, and the feet and legs are orange with brown shading. In both sexes the brilliant blue wing speculum is enlarged, preferably spilling onto the tertial feathers. Both sexes go through an eclipse molt, during which their plumage typically darkens.

Common faults. In drakes insufficient white frosting on chests, shoulders, and sides of bodies, neck collars that are narrow or do not fully encircle the neck, and light-colored tails are undesirable for exhibition. In ducks common faults are nearly white heads and necks lacking the desired fawn base color and excessively white or dark bodies lacking sufficient frosting. In both sexes dull or small wing speculums are exhibition faults (it is fairly common for older females to lack brilliance in the speculum).

Breeding hints. To improve type in the Snowy, outcross onto Grays or, better yet, Duskys with excellent conformation. To increase fawn coloring in the female Snowy, outcross onto Gray or Dusky females with the richest brown or mahogany tones. Snowy mated to Gray or Dusky normally produce all Gray or Dusky F_1 offspring. Mate the best typed F_1 males and females and they will produce offspring in the ratio of three Grays or Duskys to one Snowy. These F_2 Snowys will breed true.

Buff and Cinnamon

Buff, a soft color that does not show stains as readily as white plumage, has a plumage pattern and color genotype which includes four non *wild type* alleles — recessive *dusky*, dominant *sooty*, incompletely dominant *blue*, and sex-linked *buff*: $Pat^d Pat^d$, $So^{So} So^{So}$, $Bl^{Bl} Bl^{Bl}$, $Brsl^{bu} Brsl^{bu}$. (In Great Britain, where they consider a darker cinnamon color to be ideal, their show birds often are heterozygous for *blue*, resulting in a color often called Cinnamon in North America.) Like khaki-colored ducks, some buff ducklings hatch with varying amounts of white on the front of the neck and under the throat. Buffs lacking the *sooty* allele have more blue in their plumage (especially in the head, neck, and lower back of drakes) and overall less uniform body color in both sexes. Skin and bill color genotypes are Skn^+Skn^+, $Bil^{Sd}Bil^{Sd}$. Breeds that have a Buff variety include Call, Runner, and Orpington.

Day-old description. Buff ducklings can be mistaken for white ducklings; however, with practice and close examination, it becomes apparent that the down and bill color of buff-colored ducklings is a bit more tan than the bright yellow of white ducklings.

Adult description. The ideal color that breeders strive for in both sexes is a rich fawn buff that is uniform throughout all surface areas of the plumage. However, even the best-colored drakes usually have head, neck, lower back, undertail coverts, and speculums that are at least a shade darker or paler than the rest of the plumage. Pale-colored wings are almost universal because wing plumage is normally replaced once a year; whereas, the body plumage is replaced several times a year and therefore is not as sun-bleached as the wings.

Common faults. Because buff-colored ducks carry the *blue* allele, they are predisposed to having blue-gray shading in their plumage, especially if they lack the *sooty* allele. For exhibition avoid birds that have strong blue overtones, distinct lacing, extremely light- or dark-colored plumage, and drakes with any hint of a neck ring or claret-colored breasts.

Breeding hints. Some strains of buff-colored ducks are prone to produce females with facial stripes and drakes with claret on the chest. This problem can be eliminated by breeding from solid-headed females and drakes that had no facial stripes in their juvenile plumage.

Black

In solid black breeds, Cayugas and East Indies have the most perfected black plumage, with the best specimens having spectacular green iridescence. The plumage pattern and color genotype includes at least two non *wild type* alleles — recessive *dusky* and incompletely dominant *extended black: $Pat^d Pat^d$, $E^E E^E$*. Some blacks I have tested also carry dominant *sooty, $So^{So}So^{So}$*.

Due to genetic quirks of black plumage, some offspring will have a bit of white on the front of the neck or under the bill (even if breeding birds lacking this white have been carefully selected for many generations), and some ducklings have clubbed down, which typically is most evident on the lower back of ducklings. Skin and bill color genotypes are Skn^+Skn^+, $Bil^{Sd}Bil^{Sd}$ in individuals having greenish black bills. In those with bluish black bills, the skin and bill color genotypes are Skn^P-, $Bil^{Sd}Bil^{Sd}$. Breeds that have a Black variety include Ancona, Call, Cayuga, Crested, East Indie, Magpie, Runner, Silkie, and Swedish.

Day-old description. Ducklings have black down (often with a yellow blush on the chest and around the bill), bills, legs, and feet. Many strains of homozygous black ducks produce at least a few ducklings with clubbed down. Clubbed down is tightly curled and tends to break off as the newly hatched duckling dries, leaving bald patches. Ducklings with clubbed down typically cannot be distinguished from other ducklings once they have feathered out. The incidence of clubbed down seems to be much higher in lines of black ducks that have been selected to have a minimum of aging white plumage (see Breeding Hints below).

Adult description. Ideally, both sexes have jet black plumage with iridescent emerald sheen covering as much of the surface as possible; black bills (even the best-colored drakes usually have a bit of green or blue at the tip); and black to dusky black legs and feet (normally shaded with varying amounts of orange in mature drakes, especially as the breeding season approaches).

The green iridescence of black plumage is produced by tiny prisms on the feathers. Because the sheen is caused by refracted light and not pigment, the quality of lighting under which black ducks are viewed greatly influences their color. (The best green color is obtained when black ducks are seen in diffused light consisting predominantly of medium to short wavelengths. Long-wavelength light causes black ducks to have purple or bronze hues.) Also, black plumage that is worn loses iridescence and can look dark brown. Brilliant black plumage is restored after the bird goes through a complete molt. Black ducks that consume a diet high in yellow xanthophylls tend to have more purple in their plumage.

Common faults. For exhibition avoid birds with brown lacing or penciling (check under the wing and on the throat), prominent white (some judges check under the throat), and strong purple sheen (even the best birds often have a bit of purple barring, especially under long-wavelength light). Remember, in breeds with unusual conformation, such as the Runner, breed type is more important than color perfection.

Breeding hints. To avoid unnecessary frustration, people who raise solid black ducks need to understand that there are two types of white associated with them.

First: Juvenile white, if present, is seen in the first plumage, usually on the front of the neck or under the bill. Because the presence of juvenile white can increase from one generation to the next, it is not advisable to breed from individuals with prominent juvenile white. No matter how carefully breeders are selected, juvenile white cannot be totally eliminated from a strain of black ducks (in fact, there is good evidence that its presence is essential for hardiness and viability). In carefully bred strains 15 to 30 percent of offspring will typically display some juvenile white.

Second: Aging white is the equivalent of graying hair and is associated with green sheen. It normally shows up anywhere from a few months to a few years of age and is most prevalent in females. Aging white normally starts out as a few feathers with white edging, gradually increasing until the bird is mottled with white. Some black ducks eventually become primarily white with age, even if they were solid black show specimens as young birds. People often have difficulty with this fact, but ducks that develop aging white almost always produce offspring with greener iridescence than those individuals that stay solid black as they age.

To produce offspring with the most iridescence, choose brilliant drakes that have green on the primary flight feathers and their under-wing covert feathers, and mate them to ducks possessing good green sheen and that display at least a bit of aging white by the time they are four to twelve months of age.

To produce solid black females that can be shown as old ducks, mate drakes and ducks that have no aging white or a minimum at two years of age.

Blue and Silver

In Runners Blue is officially known as Cumberland Blue (commemorating the Runners first raised in Cumberland County, England). The plumage pattern and color genotype of Blue is the same as for Black, with the addition of a single incompletely dominant *blue* allele: Pat^dPat^d, E^EE^E, $Bl^{Bl}Bl^+$. (The genotype of Silver or Silver Splash is the same as for Black, with the addition of two incompletely dominant *blue* alleles.) Blue ducks are affected by juvenile white and aging white in a manner similar to blacks. Skin and bill color genotypes are the same as for Blacks. Breeds that have a Blue variety include Ancona, Call, Cayuga, Crested, East Indie, Magpie, Runner, and Swedish.

Day-old description. Blues have bluish gray down, dark blue or green bills, and gray legs and feet shaded with yellow. Silvers are paler.

Adult description. Blue plumage color is highly variable, even among full siblings. Ideally, it is a medium to medium-dark shade of rich blue gray with as little foreign color as possible (I have never seen a blue duck of any breed without at least a little black flecking somewhere in its plumage). Drakes typically are darker than ducks. The head and neck are darker than the rest of the plumage. Bills are green blue in drakes and nearly black in ducks. Feet and legs are gray with varying amounts of orange shading, which increases with age and during the breeding season.

The plumage is enhanced when the feathers are laced with dark blue. Lacing is most pronounced over the shoulders and on the back, and drakes normally are the most distinctly laced. Lacing fades as the feathers age.

The *blue* allele "leaks," allowing black to show through in the form of small flecks on individual feathers, entire feathers, or patches of feathers (rarely, an entire wing or side of a bird is black).

Common faults. For exhibition avoid birds with pronounced brown or pale blue color and conspicuous white feathering. Blues often develop white mottling as they age. As blue feathers are exposed to sunlight, they often develop a brown or yellowish hue — sound color is restored when old feathers are molted and new plumage is grown. A deficient diet or one containing large amounts of yellow xanthophylls, such as corn, can also negatively affect blue color.

Breeding hints. Blue drakes mated to Blue ducks produce offspring in a ratio of one Black to two Blues to one Silver. Blue × Black matings produce half Blacks and half Blues, while Blue × Silver produces half Blues and half Silvers. Silver × Black produces all Blue offspring. Breeders often debate the pros and cons of these different matings, but all four of them can produce excellent Blue specimens. The main advantage of the Blue × Blue mating is that birds with distinct lacing can be chosen for both sides of the mating. To avoid having Blue ducks that gradually become pale and washed out, it is helpful to select breeders with rich color.

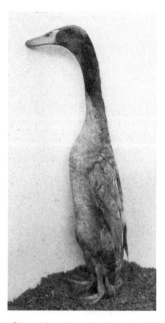

Shown here, a prize-winning Cumberland Blue Runner drake.

Chocolate, Lavender, and Lilac

Chocolate is a relatively rare color in ducks of Mallard descent and does not have the reddish tone seen in Chocolate Muscovies. The plumage pattern and color genotype of Chocolate is the same as for Black, most commonly with the addition of *sex-linked brown: Pat^{d}Pat^d, E^{E}E^E, Brsl^{br}Brsl^{br}.* There is a rare form of Chocolate that is softer colored due to *sex-linked buff: Pat^{d}Pat^d, E^{E}E^E, Brsl^{bu}-Brsl^{bu}.*) Lavenders and Lilacs have the same plumage genotype as Chocolates except that Lavenders are heterozygous for *blue* and Lilacs are homozygous for *blue*. Chocolate, Lavender, and Lilac ducks are affected by juvenile white and aging white in a manner similar to Blacks. Skin and bill color genotypes are the same as for Blacks. Breeds that have Chocolate (and rarely Lavender and Lilac) varieties include Ancona, Call, Magpie, and Runner.

Day-old description. Ducklings are dark brown (often with a yellowish blush on the chest and around the bill) with dark brown bill, legs, and feet.

Adult description. As noted under genotype, there are two forms of Chocolate in ducks of Mallard descent. In the more common form, both genders are dark brown (sometimes appearing black from a distance), often with green sheen (especially on the head, neck, back, and rump of drakes), dark green to blackish brown bills, and dark brown legs and feet that are shaded with

A pair of day-old Chocolate Anconas showing the broken pattern of their down, feet, and bills.

varying amounts of orange in sexually mature birds (especially drakes). The second form of Chocolate is rare, softer in color, and normally lacks the green sheen.

Common faults. For exhibition avoid birds with brown lacing, penciling, or obvious white mixed in their solid plumage. This color fades under prolonged exposure to sunlight, so be careful about culling Chocolates for pale or uneven color when they are not in prime feather condition.

Breeding hints. Birds that do not develop aging white in their plumage until they are older make valuable breeders. Type can be improved by outcrossing onto Blacks possessing outstanding conformation. The mating of Black drakes onto Chocolate ducks produces all Black offspring, but the sons are heterozygous Chocolate "carriers." If these heterozygous males are mated back to Blacks, half of their daughters will be Chocolates, or if they are mated to Chocolates, half of both their sons and daughters will be Chocolates. If a Chocolate drake is mated to Black ducks, it is a sex-linked mating; all sons will be Blacks that carry one Chocolate gene, and all daughters will be true breeding Chocolates.

Self White

White is the most common color in domestic ducks. When looking at White ducks, keep in mind that they have a full complement of plumage pattern and color genes. The reason they are white is that at the locus that determines whether or not plumage pigment can be produced, they are homozygous for the recessive mutation allele C^wC^w that inhibits pigment formation (occasionally, this allele will allow small amounts of pigment to "leak" through, causing colored spots of varying size).

If you are interested in finding out what patterns and colors lie hidden "under" your Whites, cross them onto homozygous Grays. Any color or pattern characteristics in the first-generation offspring other than wild type are dominant alleles carried by the White parent. To find possible recessive pattern and color genes, mate the F_1 offspring together, and hatch as many F_2 offspring as possible. You may uncover a new color!

In White ducks with yellow bills, the known skin and bill color genotype is Skn^+Skn^+. In White ducks with pink bills, the skin and bill genotype is Skn^P-. Breeds that have a White variety include Aylesbury, Bali, Call, Campbell, Crested, Hook Bill, Mallard, Pekin, Runner, and Silkie.

Day-old description. Ducklings have yellow down and bills and orange legs and feet. Pink-billed breeds such as the Aylesbury have pink bills at hatching.

A few strains produce some offspring with varying amounts of color in their down. Most of these birds will be pure white at sexual maturity.

Adult description. Depending on breed and how much dietary yellow and orange pigment these ducks consume, their plumage ranges from light canary yellow to satin white, and bills range from dark orange to yellow to pink. Legs and feet are always orange, and eyes are grayish blue. As females reach sexual maturity, it is common for more or less black or green spotting to develop in their bills due to hormonal changes.

Common faults. There are relatively few color faults in White ducks. For exhibition creamy yellow plumage is preferred in Pekins, whereas satin white is desirable in all other breeds. Black or green spotting in the bill is a show disqualification in White drakes of all breeds, while it is only a small deduction in ducks. In some White breeds females with dark color in their bills or legs are more productive layers than females with solid orange and yellow legs and bills.

Breeding hints. Normally, the only color characteristic that breeders of Whites need to pay attention to is the color of the bill and legs. However, if good productivity is more important to you than the exact color of the bill and legs, be careful about putting excessive emphasis on clean bill and leg color in sexually mature ducks.

White Bib Pattern, Dominant

I have only seen well-defined, *dominant bib* on black, blue, silver, chocolate, lavender, or lilac-colored ducks. The size and shape of bibs vary considerably, even among siblings. The foundation genotype of ducks with *dominant bib*, regardless of their color, is Pat^dPat^d, E^E-, BiD^{Bib}-. Skin and bill color genotypes vary with the breed. Breeds that have *dominant bib* varieties include Call, Orpington, Swedish, and, rarely, Runner.

In dusky-colored ducks, *dominant bib* sometimes manifests itself as a small white spot on the front of the neck or upper chest. The apparent linkage of *dusky*, *extended black*, and *dominant bib* probably explains why solid-colored varieties, such as Black Cayugas, routinely produce some offspring with more or less white on their chests, even if breeding stock with this "fault" has not been used for countless generations.

Day-old description. The down color is black, blue, silver, chocolate, lavender, or lilac, ideally with a well-defined yellow bib on the chest and a bit of white on the wingtips. Most ducklings have varying amounts of yellow on the face, especially around the bill and behind the eyes. Bills vary considerably but often are dark green or blue marked with varying amounts of yellow. Legs and feet are dark, shaded with varying amounts of yellow. With practice you can determine the number of white flights a newly hatched duckling will have on each wing by counting the primary flight down strands.

Adult description. On average drakes have larger bibs than ducks. Ideally, bibs are solid white, are symmetrically shaped, and have clearly defined borders. The exact size is not critical, but bibs that run all the way back toward the tail on the underbody are overdone, whereas a mere spot is too small. Breed standards call for the outer two or three flight feathers to be white. The exact number of white flight feathers is a relatively minor detail that should not be overemphasized by breeders or judges. Blue Swedish drakes typically have greenish blue bills, while those of ducks are dark blue or green. The legs and feet are orange shaded with varying amounts of gray (nearly black in some fully mature females).

Common faults. For exhibition avoid birds with considerable white in their faces, bibs that are broken with color, and blotchy colored bills (bill color often, though not always, darkens with age). It is common for bibbed ducks (especially females) to develop white mottling in the colored portion of their plumage as they age.

Breeding hints. Perfecting the bibbed pattern is a challenge. Choose breeders with the cleanest and most symmetrical bibs possible. Keep in mind that, usually, drakes with small bibs will produce daughters with even smaller bibs and ducks with large bibs will produce sons with huge bibs and white underbodies. The more ducklings that are hatched, the better the chances are of producing some birds with outstanding markings. With practice you will be able to identify the best-marked birds the day they hatch.

White Bib Pattern, Recessive

Well-defined white bibs on ducks with the wild Mallard or dusky plumages are due to the *recessive bib* allele, which does not have any known relationship to the *dominant bib* found in ducks having a base color of *extended black*. The base genotype can be either Pat^+Pat^+, $BiR^{bib}BiR^{bib}$ (Bibbed Mallard) or Pat^dPat^d, $BiR^{bib}BiR^{bib}$ (Bibbed Dusky Hook Bill). In Dutch Hook Bills most of the birds that we have studied have had skin and bill color genotypes of either $Skn^{P-}Skn^P$, Bil^+Bil^+, or Skn^PSkn^P, $Bil^{Sd}Bil^{Sd}$. Bibbed Mallards and the occasional Dutch Hook Bill will have *wild type* skin and bill genotype. Breeds that have *recessive bib* varieties include Dutch Hook Bill and Mallard.

Description. The white bib pattern seen in the day-old duckling is a near replica of what it will be in adulthood. Interestingly, the size and shape of the bib is distinctly different in ducks with *dusky pattern* vs. *wild type pattern*. In Duskies the white usually runs from under the bill down the front of the neck and widens on the chest, sometimes having a finger of white extending back along the underbody. Most individuals have a varying amount of white around the base of the bill and a bit behind the eyes. The number of white flights varies from none (rare) to all primaries and some secondaries.

In Mallard-colored individuals, the bib tends to be very well defined, round or oval shaped, and located on the front of the chest, with the uppermost portion coinciding with where the white neck ring would be on a Mallard drake. White in the face is rare, and the typical number of white flights varies from none to six.

Comments. *Recessive bib* is extremely rare in North America and gives breeders another interesting pattern trait to work with.

Runner Pattern

The white pattern superimposed on a colored background, as seen in Penciled and Fawn & White ducks, is due to a pair of incompletely dominant *runner pattern* genes. The plumage pattern genotype of the Penciled variety is $Pat^{d\text{-}}Pat^d$, $Brsl^{br}Brsl^{br}$, $Rn^{Rn}Rn^{Rn}$. Fawn & White and Blue Fawn Penciled have the same genotype as Penciled, except for the addition of incompletely dominant *blue*. Fawn & Whites are homozygous for *blue* ($Bl^{Bl}Bl^{Bl}$), whereas Blue Fawn Pencileds are heterozygous for *blue* ($Bl^{Bl}Bl^+$). The rare Emery Penciled has the same genotype as standard Penciled except they do not have *sex-linked brown* (Pat^dPat^d, $Rn^{Rn}Rn^{Rn}$).

While *runner pattern* can be added to any color, the most clearly defined pattern is achieved in conjunction with *dusky* as in the Penciled and Fawn & White varieties. Although rare, *runner pattern* has been combined with Mallard-colored plumage (Pat^+Pat^+, $Rn^{Rn}Rn^{Rn}$), resulting in a pretty variety which is called Pied in Great Britain. It is especially difficult to obtain a cleanly marked pattern when *runner pattern* is combined with *extended black*. Ducks that are heterozygous for the *runner pattern* allele typically are solid colored except for a few white flights and a small white patch on the front of their necks. Skin and bill genotypes vary with the breed and variety.

Breeds that have Penciled, Blue Fawn Penciled, Emery Penciled, Fawn & White, or Pied varieties include Call and Runner.

Day-old description. The down pattern of day-olds is a replica of the adult pattern. Penciled ducklings are brown and yellow. In Fawn & Whites the fawn portions of the down are only slightly darker than the yellow portions.

Adult description. Ideally, the white areas include the upper two-thirds of the neck, the throat, a finger that extends from the back of the head to and partially encircles the eye, a line that divides the bill from the cheek patches, a belt across the underbody and the outer two-thirds to three-quarters of the wings.

Common faults. The runner pattern (along with the magpie) is the most challenging to perfect. Therefore, breeders and judges should not overemphasize minor details but rather should focus on overall

A champion Penciled Runner drake with outstanding markings

clarity and symmetry. Common faults include excessively large or small head caps and cheek markings, head and cheek markings that trail down onto the neck, colored patches on the throat and neck, "snow" flecks on the shoulders and back, lack of a white belt across the underbody, colored wing feathers, and solid white tail feathers.

Breeding hints. Drakes with small cheek patches tend to produce daughters with medium-size cheek patches, whereas ducks with excessively large cheek patches tend to produce sons with medium-size patches. When possible, use breeders with clearly defined markings. However, poorly marked birds that are out of a good strain can produce some excellent marked offspring.

Keep in mind that the expressivity of *runner pattern* is highly variable. Therefore, no matter how carefully breeding stock is selected, there will always be variability in the markings of their offspring. The more ducklings you hatch, the better the chances of producing some birds with outstanding markings. With practice the best-marked individuals can be identified the day they hatch. Breeders who produce well-marked runner pattern ducks have perfected one of the two most challenging patterns in ducks.

Magpie Pattern

The bold magpie pattern results from combining *extended black*, *runner pattern*, and *dominant bib* genes. This genotype is highly variable in its expression, resulting in wide variation in the plumage pattern of even the most carefully bred strains of magpie-marked ducks. The plumage pattern and color genotype of Black Magpie is Pat^dPat^d, E^EE^E, $Rn^{Rn}Rn^{Rn}$, $BiD^{Bib}BiD^{Bib}$. Blue Magpie is the same as for Black except for the addition of a single *blue* allele; Silvers are homozygous for *blue*. Chocolate Magpie is the same as Black except for the addition of *sex-linked brown*. Skin and bill color genotypes are Skn^+Skn^+, $Bil^{Sd}Bil^{Sd}$ but due to the interaction of the white pattern genes (*dominant bib* and *runner pattern*), the bills typically are spotted with varying amounts of yellow or orange. Breeds that have a Magpie variety include Call and Magpie.

Day-old description. The down pattern of the day-olds is a replica of the adult pattern. Their bills, legs, and feet are yellow, sometimes with black or gray markings. (Some ducklings hatch out solid yellow and will be white at maturity.)

Adult description. Ideally, the crown of the head and the back mantle (including the shoulders, back, and tail) are colored, with the rest of the plumage being white. The bills are yellow or orange with green shading in young birds, gradually darkening with age. The legs and feet are orange-red shaded with grayish black, darkening with age, especially in the females.

Common faults. The magpie pattern (along with the runner pattern) is the most challenging pattern to perfect in ducks. Therefore, both breeders and judges should not place undue emphasis on minor details but rather should emphasize overall clarity and symmetry of the pattern. Common faults include head caps that cover less than half the crown or are totally missing (the colored caps gradually disappear on most females as they age), caps running down the back of the neck or spilling below the eyes, color on the breast or sides of the body, and white on the shoulders, back, and tail. With age the colored portions of the plumage will often gradually turn white, especially in females.

Breeding hints. White tends to increase in drakes and color increases in ducks from one generation to the next. When a standard-marked drake is mated to a standard-marked duck, they tend to produce excessively white sons but well-marked daughters (and a fair number of solid white offspring). In general, females with excessive color (including color under the eyes) produce the best-marked sons.

If they are descended from a good strain, even poorly marked Magpies can produce some offspring with excellent markings. The more ducklings that are hatched, the better the chances of producing some outstandingly marked birds. With practice the best-marked birds can be identified the day of hatching. Like breeders of runner pattern ducks, breeders who produce well-marked magpie pattern ducks should be satisfied knowing they have perfected one of the top two most challenging duck colors.

Ancona (Broken) Pattern

The broken ancona pattern, as is the case with the magpie pattern, results from combining *extended black, runner pattern*, and *dominant bib* genes. However, in Anconas, the best- and wildest-marked individuals are heterozygous for *extended black* ($E^E E^+$), whereas Magpies are always homozygous ($E^E E^E$). Selective breeding has produced the crazy-quilt effect of the ancona pattern. The plumage pattern and color genotype of the best broken pattern Black Ancona is $E^E E^+$, $Rn^{Rn} Rn^{Rn}$, $BiD^{Bib} BiD^{Bib}$. Blue Anconas have the same genotype as the Black, except that they are heterozygous for *blue*; Silvers are homozygous for *blue*. Chocolate has the same genotype as Black, with the addition of *sex-linked brown* (very rarely *sex-linked buff*). Lavender Anconas have the same genotype as Chocolate, except they are heterozygous for *blue*, while Lilacs are homozygous for *blue*.

Some Anconas carry *dusky pattern* whereas others are *wild pattern*. When two Anconas that are heterozygous for *extended black* are bred together, approximately 25 percent of their offspring will not carry *extended black*. Anconas without *extended black* can have a wide variety of plumage colors and patterns (including Fawn & White, Blue Fawn Penciled, Penciled, Emery Penciled, and Pied) and are often grouped together under the label of Tri-Colored Anconas. Skin and bill color genotypes are $Skn^+ Skn^+$, $Bil^{Sd} Bil^{Sd}$, but due to the interaction of the white pattern genes (*dominant bib* and *runner pattern*) the bills typically are spotted with varying amounts of yellow or orange. Breeds that carry the broken pattern include Ancona and Call.

Day-old description. The pattern of the day-olds is a near replica of the adult plumage (some ducklings hatch out solid yellow and will be white at maturity). The bill, legs, and feet are yellow or orange, usually marked with more or less black or dark brown.

Adult description. There is no set pattern. In fact, the more haphazard and asymmetrical the markings, the better. Ideally, there should be bold patches of color under the eyes, on the chest, on the sides of the body, and on the back. A dramatic pattern is paramount. Because the best-marked individuals carry a single *extended black* allele, Blacks and Blues (especially the drakes) often have reddish brown shading in the colored portions of their plumage. This "rust" is considered a fault in most Black and Blue ducks, but is a necessary characteristic of the broken pattern. The bill, legs, and feet should be as spotted and blotchy as possible in young mature birds, with these extremities turning darker and often more solid with age.

Common faults. The most common pattern faults include no color patches below the eyes or on the chest, sides of body, or back or having distinct magpie markings.

Breeding hints. To minimize the number of solid white offspring produced, avoid mating two birds that are more than two-thirds white. Ancona pattern ducks often produce some offspring with runner or magpie patterns. If these are mated back to broken-pattern birds, they will produce some offspring with good broken pattern. Better yet: mate an Ancona with tri-colored runner pattern plumage to an Ancona with magpie pattern plumage; this typically results in 100 percent of the offspring having bold, ancona markings.

To maximize the genetic diversity of this rare breed, it is advisable to not separate out the different color varieties for breeding pens. Rather, interbreed the various colors together. This practice can reduce the effects of inbreeding depression.

Muscovy Colors

The APA currently recognizes only four color varieties of Muscovies: White, Black, Blue, and Chocolate. However, more than two dozen varieties — some of which are quite spectacular — are raised in North America. Except for the Loonie variety, I have raised and studied all the colors described here. These colors and patterns are the result of various combinations of *wild type* alleles plus one or more of the nine main mutations, as outlined in the table on the following page.

Muscovies naturally have pinkish white skin. If their diet contains sufficient quantities of yellow pigment, their skin can have a yellow cast.

PLUMAGE COLOR AND PATTERN ALLELES
in Muscovy Ducks

Name	Symbol	Relationship to "Wild"	Main Visual Effects
Wild Pattern	Pat^+	Wild type	Wild Muscovy pattern
Dusky Pattern	Pat^d	Recessive	Extends dark pigment in ducklings; reduces brown markings in adult
Blue	Bl^{Bl}	Incompletely dominant	Dilutes black pigment to blue gray in single dose, to silver in double dose
Wild	Bl^+	Wild type	Normal plumage color expression
Wild	$Brsl^+$	Wild type	Normal plumage color expression
Brown	$Brsl^{br}$	Sex-linked recessive	Changes black pigment to reddish brown
Wild	L^+	Wild type	Normal plumage color expression
Lavender	L^{la}	Recessive	Dilutes black pigment to uniform lavender
White	P^W	Incompletely dominant	Prevents pigmentation of plumage
Wild	P^+	Wild type	Normal plumage pigmentation
Wild	Bar^+	Wild type	Normal plumage color expression
Barred	Bar^{bar}	Recessive	Lightens duckling color; causes more or less barring in juvenile feathers
Wild	Rip^+	Wild type	Normal plumage color expression
Rippled	Rip^{rip}	Recessive	Lightens ground color; causes irregular crosshatches on feathers
Wild	Mg^+	Wild type	Normal pigmentation throughout
Magpie	Mg^{mg}	Recessive	Inhibits pigmentation on neck, wings, and underbody
White Head	WH^{WH}	Dominant	Inhibits pigmentation on head and neck
Wild	WH^+	Wild type	Normal pigmentation throughout plumage

Wild Type

This is the original color found in wild Muscovies and is sometimes seen in feral populations as well as in some domestic flocks. All other varieties are mutations of the wild type. In mature birds the phenotype of Black Muscovies is hard to distinguish from those with wild-type color.

Day-old description. Ducklings have the classic dark and yellow camouflage pattern. Compared with Mallards, the light portions of the down are paler yellow, the dark facial stripe is abbreviated and does not extend between the eye and the bill, and the dark portions of the head and neck have a browner cast. The bill is black, and the legs and feet are dusky yellow with black shading.

Adult description. The plumage is black, with considerable green and purple iridescence. The breast, sides, and underbody often have bronze overtones, especially noticeable in younger birds (overall, the feathers of the juvenile plumage are less iridescent than in fully mature birds). The wings start out solid colored, with the forewings gradually turning white over the course of several years. The facial skin patch is normally smooth and pigmented with considerable black.

Common faults. There are few domestic Muscovies with good wild-type color. The most common faults are the presence of excess white in the body plumage and the lack of black pigment in the facial skin patch.

Breeding hints. There is no written standard for the wild-type color. However, using wild birds as the standard, select for breeding those specimens that have the classic dark and yellow camouflage pattern as ducklings; display brown or bronze shading in their juvenile plumage; and as adults exhibit brilliant green and purple iridescence on the surface of their feathers and have as much black as possible on their facial skin patches.

Black

From 1904 until 1998 this variety was officially known as Colored. Genetically, Blacks are *wild type* except for two recessive *dusky* alleles, Pat^dPat^d. Most Muscovies carrying *dusky* will have at least a small white patch at the juncture of the bill and throat.

Day-old description. Ducklings are solid black (sometimes with a spot of yellow on the front of the neck) with brown shading, especially on the head and neck. The bill, legs, and feet are black to dusky black.

Adult description. The plumage is black with very striking green and purple iridescence (the iridescence is less pronounced in the juvenile plumage). The wings usually start out solid colored, with the forewings gradually turning white over the course of several years. With age more or less white flecking often develops in the head and neck plumage. The bill is pink, shaded with varying amounts of black. Legs and feet are dusky yellow to black.

Common faults. For exhibition, faults to avoid include pronounced black in the facial caruncles (even the best drakes normally have a bit of black near the eyes), excessive white in the plumage other than the forewings, lack of white forewings in mature specimens, and pronounced brown in the plumage of mature birds (most mature "Blacks" with brown in their plumage are actually wild type). A white spot on the throat at the juncture of the bill is not a fault.

Breeding hints. In general, individuals with the blackest feet and legs have more black or mulberry in their faces. Therefore, it is usually helpful to select breeders with dusky yellow feet and legs. Because the white flecking of the head and neck plumage often does not develop until a bird is six to eighteen months old, two-year-old birds with the least white in these areas make valuable breeders.

Blue and Silver

The plumage pattern and color genotype of exhibition Blues is homozygous for recessive *dusky* and heterozygous for incompletely dominant *blue*, Pat^{d}-Pat^{d}, $Bl^{Bl}Bl^{+}$. Some nonexhibition Blues are *wild type* at the *Pattern* locus, Pat^{+}-. Silvers are homozygous for *blue*, $Bl^{Bl}Bl^{Bl}$. The bill is pink, shaded with varying amounts of black. Legs and feet are dusky yellow to black.

Day-old description. In Blue ducklings the black portions of the down are diluted to bluish gray and the bill, legs, and feet are slightly lighter than those of Blacks. Silver ducklings are pale silver.

A juvenile Silver Muscovy duck, bred on our farm, will produce 100 percent Blue offspring when bred to a Black drake.

Adult description. The plumage of Blues is an attractive bluish gray, ideally with each feather of the back and sides of the body laced with a darker border. Often drakes are darker colored with more pronounced lacing than ducks (the juvenile plumage is duller). Everything else is similar to Blacks. Silvers vary from a light to a medium shade of silver.

Common faults. Even full siblings can vary considerably in the shade of their plumage. The exact shade of blue gray is not critical, although medium to medium-dark birds are generally preferred for exhibition. Black tends to "leak" through to produce "ink spots" that range in size from a small streak to patches of solid black feathers. Blues and Silvers can have the same faults as Blacks.

Breeding hints. Excellent-colored Blues can be produced by three matings. Blue × Blue produces ducklings in a ratio of one Black to two Blues to one Silver. Blue × Black yields half Blues and half Blacks. Silver × Black gives all Blue offspring.

Chocolate

The plumage pattern and color genotype of Chocolates is the same as Wild Type or Black (the latter preferred for exhibition) plus *sex-linked brown, $Brsl^{br}$ $Brsl^{br}$*. The bill is pink, shaded with varying amounts of dark brown. Legs and feet are brownish yellow to brown.

Day-old description. Chocolate ducklings are a rich reddish brown with dark brown bills, legs, and feet.

Adult description. The plumage is a rich reddish brown with purple, and in the right light a greenish iridescence is detectable (the juvenile plumage is duller). The white markings are the same as in Blacks. Because all black pigment is diluted to brown, Chocolates do not have true black in their faces.

Common faults. As the season progresses Chocolates often display faded plumage, usually the result of exposure to sunlight and everyday wear and tear. A well-balanced diet and shade help Chocolates stay in good feather condition longer.

Breeding hints. Type and size can be improved by mating Chocolates to outstanding Blacks. A Chocolate drake mated to a Black duck produces all Black sons (that are Chocolate carriers) and Chocolate daughters. The reciprocal cross of Black drake to Chocolate duck produces all Black offspring, with the drakes being Chocolate carriers.

White

White plumage in Muscovies is caused by an incompletely dominant *white* allele at the *Pigment* locus, P^WP^W, whereas in Mallard-derived ducks, white plumage is the result of a recessive *white* allele at the *Color* locus, C^wC^w. Muscovies that are heterozygous for *white* are haphazardly splashed with white and color. The bill of White Muscovies is pink to pinkish white, and the legs and feet are yellow or orange.

Day-old description. Ducklings are lemon yellow with pink bills and yellow legs and feet. Sometimes they have a colored spot on the top of their heads.

Adult description. Most adults are pure white, even if they possessed a colored head spot in their juvenile feathering.

Common faults. In first-year birds color on top of the head is not a disqualification for exhibition. Because of the genotype of White Muscovies, drakes are prone to developing dark brown or black markings in their bills, which unfortunately is a show disqualification. Breeders who attempt to eliminate dark pigment in the bills of drakes often report reduced fertility.

Breeding hints. If you are intent on reducing the incidence of dark pigment in the bills of drakes, it can be helpful to choose clean-billed breeding drakes that are at least two years old and mate them to females with no black or brown in their bills. However, keep in mind that you may be inadvertently reducing the fertility of your strain.

Nonstandard Varieties

The following varieties have not been recognized by the APA. All of them are attractive, and several are common in some parts of the world.

Lavender

This lovely, rare color is sometimes called Self Blue because each feather is uniformly colored throughout its surface. To confuse matters further, some people call this variety Silver, despite the fact that the phenotype and genotype of Lavender and Silver are distinctly different. True Silvers are lighter in color, lack the purplish hue of Lavenders, and genotypically are Black or Wild Type with two incompletely dominant *blue* alleles, $Bl^{Bl}Bl^{Bl}$. On the other hand, Lavenders are Black or Wild Type with two recessive *lavender* alleles, $L^{la}L^{la}$.

Day-old description. Lavender ducklings are medium bluish silver, a shade or two darker than true Silvers.

Adult description. The plumage is a uniform bluish lavender without darker lacing on the perimeter of the feathers or black ink spots. The white markings are the same as in Blacks. The bill is pinkish, often with a dark saddle, and the legs and feet are gray.

Common faults. Lavenders can have the same faults as Blacks. This color will eventually fade and discolor when exposed to bright sunlight. The true color is restored following the molt.

Breeding hints. The best-colored Lavenders have a Black genetic base rather than Wild Type. To improve the size and conformation of Lavenders, outcross them onto the best Blacks obtainable. The first generation offspring will be black (if the Blacks were homozygous), but when the F_1s are intermated, they will produce F_2 offspring in a ratio of three Blacks to one Lavender.

Barred Pattern

Barred Muscovies are regionally common, and with practice are easily differentiated from Rippled Muscovies, with whom they are often confused. The barred genotype is homozygous for recessive *barred*, $Bar^{bar}Bar^{bar}$, and is a pattern that can be superimposed on any color, including Black, Blue, Blue Fawn, Chocolate, Lavender, Lilac, Pastel, and Silver. *Barred* allele can also be combined with *magpie*, *rippled*, and *white head* alleles.

Day-old description. The unique down of Barred Muscovy ducklings is diagnostic and resembles that of Harlequins, with their yellow down tipped with a bit of color. The bills, legs, and feet are a shade or two lighter than in nonbarred individuals.

Adult description. Barring is fairly distinct in the feathers of most juveniles but largely disappears in the adult plumage, where it sometimes causes a marbled effect, especially on the sides and underbody. With careful selection of breeding stock, barring can be enhanced somewhat in the adult plumage. In some countries Barreds are preferred for commercial production because of their lighter undercolor.

Common faults. Barreds can have the same faults as Blacks.

Breeding hints. Size and type can be improved by outcrossing Barreds to the best Blacks, Blues, Chocolates, or Lavenders available. The F_1 generation will be nonbarred, but one out of four of the F_2 offspring will be Barred. The Barred offspring are easily identified at hatching by their nearly yellow down.

Rippled Pattern

This distinctive pattern is rare in most localities. The typical genotype is $Pat^d Pat^d$, $Rip^{rip}Rip^{rip}$. *Rippled* can also be combined with other colors such as *blue* (Pat^dPat^d, $Bl^{Bl}Bl^{Bl}$, $Rip^{rip}Rip^{rip}$), *chocolate* (Pat^dPat^d, $Brsl^{br}Brsl^{br}$, $Rip^{rip}Rip^{rip}$ –this genotype is historically called Buff), or *lavender* (Pat^dPat^d, $L^{la}L^{la}$, $Rip^{rip}Rip^{rip}$).

People sometimes mistakenly identify Rippled as Barred, but the color of the hatchlings is diagnostic. Barred ducklings are yellow with the very tips of the down having a bit of color while Rippled ducklings of the genotype $Pat^d Pat^d$ $Rip^{rip}Rip^{rip}$ are very dark brown.

Day-old description. At first glance these can be mistaken for Blacks. However, upon closer examination, it is evident that the body, bill, legs, and feet are browner than in Blacks.

Adult description. The base color is blue gray, irregularly marked with darker crosshatches of dull black. Some birds have vertical center marks on their feathers (especially over the shoulders and back). Upon seeing this variety for the first time, people sometimes assume they are odd-colored Blues. They typically have white forewings, as in Blacks.

Common faults. Rippled Muscovies can have the same faults as Blacks.

A Rippled Muscovy old drake with excellent markings (bred on our Waterfowl Farm).

Breeding hints. Size and type can be improved by using the same methods as with Barreds. The Rippled F₂ offspring will be slightly browner in down color than their nonrippled siblings.

Magpie Pattern

Muscovies with bold white markings are common in many countries. Birds of this phenotype go by various names, such as Duclair, Piebald, Colored and White, Parti-colored, and Magpie. The plumage pattern genotype is homozygous for *magpie*, $Mg^{mg}Mg^{mg}$. The magpie pattern can be bred into Muscovies of any color.

Description. While the *magpie* allele is highly variable in its expression, the best-marked Magpie Muscovies have well-defined colored and white sections that include a colored mantle covering the shoulders, back, and tail. They may or may not have color on the crown of the head. Muscovies that are haphazardly splashed or mottled with white usually are not Magpies but rather are heterozygous for incompletely dominant *white*, P^W-.

Common faults. In Great Britain magpie-patterned Muscovies are standard varieties and are called Black-and-White, Blue-and-White, and Chocolate-and-White. Common faults of these are poorly defined shoulder and back mantles, color spilling down on the underside of the body, and excessive color on the head.

Breeding hints. By carefully selecting breeders for several generations that possess the preferred markings, pattern uniformity can be improved in the offspring.

White Head Pattern

This unique pattern is common in some feral populations and farm flocks. The genotype can either be heterozygous or homozygous for the dominant *white head* allele, WH^{WH}-. The *white head* pattern can be combined with *black, blue, chocolate,* and *lavender.*

Description. In juveniles the head and neck are colored, gradually turning white with age. In adults the head and neck are white (often peppered with dark feathers), with the rest of the body colored like other Muscovies. Especially in some drakes, the white may spill down onto the chest and underbody.

Common faults. Keep in mind that it can take a year or more for the white head to fully develop, and ducks tend to have better markings than drakes. White in the body plumage is a common fault.

Breeding hints. Females that retain colored flecking in their head and neck plumage often produce sons with the best white-head markings. On the other hand, females with pure white heads and upper necks usually produce the best-marked daughters.

Other Varieties

Each of the following varieties results from combining two or more of the colors or patterns previously described.

Buff

These are the result of combining *sex-linked brown* and recessive *rippled*, $Brsl^{br}Brsl^{br}$, $Rip^{rip}Rip^{rip}$. Adults are a medium shade of buffish brown, with white forewings.

Blue Fawn

These are the result of combining a single *blue* allele with *sex-linked brown* $Bl^{Bl}Bl^{+}$, $Brsl^{br}Brsl^{br}$. Adults are bluish brown with white forewings.

Lilac

These are the result of combining two incompletely dominant *blue* alleles with *sex-linked brown*, $Bl^{Bl}Bl^{Bl}$ $Brsl^{br}Brsl^{br}$. Adults are pale lilac with white forewings.

Pastel

These are the result of combining *sex-linked brown* and two recessive *lavender* alleles, $Brsl^{br}Brsl^{br}$, $L^{la}L^{la}$. The adults resemble Lilacs but are slightly richer in color.

Loonie

These are the result of combining recessive *dusky pattern*, recessive *barred*, and recessive *rippled*, $Pat^{d}Pat^{d}$, $Bar^{bar}Bar^{bar}$, $Rip^{rip}Rip^{rip}$. I have not bred them, but they were produced and named by the fine geneticist W. F. Hollander, who described them as resembling the color of the Common Loon.

11

Acquiring Stock

HAVING SELECTED THE BREED or breeds you want to raise, the next step is locating suitable stock. The importance of starting with good-quality birds cannot be overemphasized. The productivity, growth rate, and size of ducks within the same breed vary a good deal, depending on the source of your stock. If your duck project is going to be economically practical and free of unnecessary problems, healthy and productive birds are essential. Remember: Poor-quality birds eat just as much as — or more than — good-quality stock. Often, cheap birds end up being more costly over the long haul.

Production-Bred Stock vs. Standard-Bred Stock

Egg- and meat-producing characteristics are given first priority in production-bred ducks, with less concern for perfect color or shape. On the other hand, standard-bred birds are used for showing in competition and are painstakingly selected for color, size, and shape, with production abilities often given second priority. In some breeds of ducks, the productivity and growth rate of exhibition strains are equal to or better than the commercial stock that is commonly sold by hatcheries. Normally, standard-bred stock is priced higher than production-bred stock.

Options for Acquiring Stock

Depending on availability and your circumstances and preferences, there can be a number of options for starting your duck flock. The three common ways to get started are with hatching eggs, ducklings, or mature stock, with ducklings being the most readily available choice in many breeds. In some breeds, only one or two of these options may be possible.

Hatching Eggs

If a dependable setting hen or incubator is available, you may wish to buy hatching eggs to start your flock. Some advantages of this method are that hatching eggs normally sell for one-third to one-half the price of day-old ducklings, and you get to experience the fun of waiting for and witnessing the hatch. Some disadvantages are that eggs vary in their fertility, they may be broken or internally damaged when shipped, and hatchability can vary widely.

If you receive a shipment of hatching eggs that is insured or you are paying COD, open the package in the presence of your postal carrier to check for breakage and to count the number of eggs received. If a substantial number of eggs are broken or there are fewer eggs than you paid for, the postal carrier will provide a claim report.

Some people feel that better hatching results are obtained if shipped eggs are "rested" for 6 to 12 hours at 55 to 65°F (13 to 18°C) prior to incubation. If you know that the eggs are older than 10 days, I recommend setting them within a couple of hours of receiving them.

Day-Old Ducklings

Purchasing day-old ducklings is the most popular method of starting a duck flock. They are more widely available and weigh less than either hatching eggs or adult stock. Ducklings are sturdy and can be shipped thousands of miles successfully if the shipper knows what he or she is doing. In our experience of mailing thousands of shipments via the United States Postal Service (and a fewer number by air freight), all the ducklings arrive safely in the vast majority of shipments. One reason why we have such a high success rate is because we ship the ducklings as soon after hatching as possible—usually within 12 hours after they emerge from the shell. As with anything, there is an element of risk, but the survival rate during shipment normally is significantly higher than for the same period of time for naturally brooded ducklings in the wild.

By the end of the incubation period, approximately two-thirds of the yolk is left unused. Shortly before the hatchling emerges from the egg, the remaining nutrient-rich yolk is absorbed into its abdomen. Thus when ducklings hatch, they do not eat or drink anything for several days while they are living off the yolk, making this the very best time to ship them.

For their safety and well-being, there are situations when ducklings should not be shipped. Ducklings from the tiniest breeds — Calls and East Indies — should not be shipped under most circumstances due to their extremely small size and accelerated metabolic rate. Due to their unique physiological characteristics, Muscovy ducklings should not be shipped unless they can be guaranteed to arrive within 48 hours after emerging from their eggs. Once ducklings have been fed and watered, they should not be shipped at all because it is extremely stressful for them and can cause permanent damage.

Ducklings are sold either sexed or straight run (i.e., the sex ratio that nature provides). Over the course of a hatching season, the ratio of males to females is approximately 50:50. However, in any given week, the sex ratio can vary widely, and you can end up with a skewed ratio simply by chance.

When ordering ducklings, give your telephone number and instruct the shipper to include it on the shipping label. If you live on a rural route, ask your postmaster to hold the ducklings at the post office and phone you upon their arrival so you can pick them up promptly. When a shipment is received, open the box in the presence of the postal employee, check the condition of the ducklings, and count the live birds. If you receive fewer live ducklings than you paid for, the postal carrier should provide a claim report.

Care of Shipped Ducklings

The first 24 hours after ducklings arrive is critical. The babies should be provided lukewarm drinking water in a suitable waterer that does not allow them to get soaked, appropriate food, and be allowed to rest in a preheated brooding area as soon as possible.

Immediately upon arrival, take the little ones from the shipping box and place them in the warm brooder, dipping each of their bills in the lukewarm water to which 1 teaspoon of honey or corn syrup per quart of water has been added. Sprinkling finely chopped lettuce, dandelion greens, or tender, young grass on the water attracts them to the water and mimics the duck weed that is the first food of wild ducklings. Letting them drink water and eat greens for 30 to 45 minutes prior to giving them prepared duckling starter will help reduce what I call "foamy mouth syndrome" — the result of ducklings eating too much grain-based feed too quickly.

Ducklings should be checked frequently the first day, but do not handle or disturb them more than absolutely necessary. Always use waterers that ducklings can drink from easily but cannot get into and become soaked.

Day-old Australian Spotted ducklings, a few hours after being taken out of the hatcher.

Buying Mature Stock

The quickest way to obtain a producing duck flock is to purchase mature birds. Poultry farms and hobbyists sometimes have adult stock available. Waterfowl adapt quickly to new climates and are readily shipped, so you can order from out-of-area breeders if the birds you want are not available locally. Healthy adult ducks ship well without food or water over long distances if the following guidelines are followed: they are shipped during cool or cold weather (preferably 75°F [24°C] or lower); they are well hydrated prior to shipping; and the shipping containers are well ventilated.

Some good places to look for duck sources include feed stores; agricultural fairs; agriculture Extension services; classified ad sections of poultry, farm, and gardening magazines and local newspapers; and appendix F (Duck Breeders and Hatchery Guide) in this book.

How Many?

The ideal number of ducks depends on your purpose for raising them, the breed raised, environmental conditions, and your management.

To estimate the number of ducks needed for a laying flock, calculate the total number of eggs desired over a year. Divide this number by the average number of eggs a duck of the breed you are going to raise will lay yearly, then add 10 percent to allow for the occasional poor layer or mortality. Keep in mind that the figures in the Breed Profiles chart (pages 28–29) for the yearly egg production of the various breeds are for ducks that are fed concentrated feeds and exposed to no less than 14 hours of light daily during the laying season.

When purchasing breeding stock as day-olds, plan on culling 10 to 50 percent of the inferior birds at maturity. Remember, no matter how carefully bred, not all offspring are suitable for breeding purposes.

When purchasing show birds as day-olds, plan on raising a minimum of two or three ducklings for every show bird desired. In general, the more ducklings raised, the better your chances of raising an elite show bird.

GOTCHA!

When Mr. DeFord told me I could have any of the ducks I caught in his gravel pit, he must have thought the chances were slim that a twelve-year-old could capture those half-wild Mallards. The first step in my plan was to shamelessly bribe the ducks with wheat every day after school. I then constructed a C-shaped trap from 4-foot-high chicken wire and metal fence posts pounded into the ground. After a week of baiting, I put down the feed in the back of the enclosure and hid behind a nearby log. The ducks rushed in to grab the feed as soon as I was out of sight. Moving quickly, I dashed to the front of the trap and closed it with an attached flap of wire. Within an hour, the flock of surprised broadbills had a new home in my duck yards.

12

Incubation

FOR MANY OF US WHO OWN POULTRY, the incubation and hatching of eggs is one of the most fascinating phases of raising birds. When holding an egg in one's hand, it is difficult to comprehend that inside the shell there exists every element necessary for the beginning and growth of a new life. In fact, when a fertile egg is laid, an embryo several thousand cells in size has already formed. If stimulated by warmth and movement, that tiny spark of life will grow, break from the shell, and present itself as part of a new generation of ducks!

Hatching Eggs

Any egg that is fertile has the potential to hatch. However, for consistently good results, hatching eggs need to be produced by healthy ducks that live in a good environment, are not obese, and consume an adequate diet. In general, breeding feeds have higher levels of protein and most vitamins when compared to layer rations. If you cannot find a duck or waterfowl breeder feed in your local feed stores, a game-bird breeder feed will normally work well (see chapter 15, Understanding Feeds).

Once an egg that is going to be incubated is laid, the use of proper handling, storing, and incubation procedures will increase its chances of producing a viable duckling.

Gathering

Eggs that are going to be incubated by a foster hen or in an incubator should be gathered at least twice daily to protect them from predators and prolonged exposure to the elements. Hatching eggs must always be handled gently so that the diminutive embryo is not injured or the protective shell cracked. Do not roll eggs over and over, jolt them sharply, or handle them with dirty hands — all of these can destroy fertility.

Cleaning

Eggs that are nest-clean — do not have soil, mud, or fecal material adhering to the shell — typically hatch the best. In reality, if you have more than a few ducks in a pen it can be challenging to gather eggs clean enough that washing is unnecessary. Usually, it is best if dirty eggs are washed as soon after gathering as possible. Because washing eggs increases the rate at which they dehydrate during incubation, in many situations if any of the eggs need to be washed, it is advisable to wash them all.

Washing does have possible negative effects on duck eggs. It removes the cuticle (a protective film on the shell that reduces dehydration and helps screen out pollutants), which often results in the need to raise the humidity level (usually by 2 to 5 degrees on the wet bulb thermometer) during incubation. Nonetheless, it is preferable to wash dirty eggs rather than set them uncleaned, since soiled eggs create an unsanitary condition under the female or in the incubator and may explode during incubation due to the buildup of pressure caused by harmful gases produced by bacteria in contaminated eggs.

When eggs are washed, always use clean water that is 10 to 25°F (5.5 to 14°C) warmer than the eggs. Adding an antibacterial soap or dish detergent to the wash water and/or a hatching-egg disinfectant to the rinse water is beneficial in most situations. Washing with dirty water spreads contaminants from egg to egg, while washing with cold water forces filth deeper into the shell pores.

Selecting

Not all fertile eggs are suitable for hatching. Those used for setting should have normal shells and be average to large in size. Extremely large eggs often have double yolks and seldom hatch. Eggs having irregular characteristics or cracks should not be set. Valuable eggs with small, tight cracks can sometimes be saved by placing a piece of masking tape over the fracture after the egg has been cleaned and dried.

Storing

Proper care of eggs prior to setting is just as important as correct incubation procedures. In working with owners of small poultry flocks, I have found that careless handling of eggs before incubation is one of the leading causes of poor hatches. Always keep in mind that no matter how faithful a setting hen is or how carefully the incubator is regulated, a poor hatch will result if the embryos have been weakened or killed during the holding period.

Where. Eggs should be stored away from direct sunlight in a cool, humid location. We have had good success storing our hatching eggs on egg flats placed inside cardboard boxes that are then placed in large plastic bags that are closed at the top. These boxes are kept in an unheated basement room. Egg cartons can be used for small quantities of eggs. Due to uniform temperature and typically higher humidity, cellars and unheated basements often are good storage locations. Refrigerators are usually colder than ideal.

Position. The position in which eggs are stored prior to incubation has relatively little effect on hatchability. The results of a study involving thousands of duck and goose eggs showed minimal difference in the hatchability of eggs stored on their sides, vertical with the air cell up, or vertical with the air cell down. However, when eggs are stored in 12-egg cartons or 30-egg flats (20-egg turkey flats are best for large duck eggs), less breakage occurs if the eggs are positioned with their large end up.

Temperature. The ideal storage temperature for hatching eggs that are held for 10 days or less seems to be 55 to 65°F (13 to 18°C). If eggs are kept for a longer time, a temperature of 48 to 52°F (9 to 11°C) will produce better hatches. Because wide temperature fluctuations reduce the vitality of embryos, it is wise to store eggs where the temperature stays at a fairly constant level. In a study of the effects of storage temperature on hatchability in duck eggs that were stored for 10 to 14 days, the following results were obtained:

Storage Temperature: °F (°C)	Hatchability Percentage
38–40 (3.5–4.4)	61
60–62 (15.5–16.7)	73
76–82 (24.4–27.8)	42
Storage Period	
1–7 days	71
8–14 days	64
15–21 days	47
22–28 days	18

Turning. Duck eggs that are held for 5 days or less show little improvement in hatchability when turned during the storage period. On the other hand, when eggs are stored longer than 5 days, the hatch rate can be increased 3 to 15 percent by turning them daily while they're being saved for incubation. If eggs are stored in egg cartons or flats, they can be turned by leaning one end of the container against a wall or on a block at an angle of 30 to 40 degrees each day, alternating the end that is raised.

Storage period. Ordinarily, the shorter the storage period, the better the hatch. A few eggs that have been held 4 weeks or longer may hatch, but for consistently good results the general rule is to keep eggs for no more than 7 to 10 days before setting them. Each day of storage adds approximately 1 hour to the incubation period. There is some indication that eggs hatch better if incubation does not commence until at least 6 hours after they are laid. The negative effect of long storage periods on hatchability can be seen in the results obtained from a test involving several thousand eggs (see chart at left).

Incubation Period

The normal incubation period for ducks derived from Mallards varies from 26 to 29 days, depending on the breed or strain. Muscovies require approximately a week longer — 33 to 35 days. High temperatures and low humidity during storage and/or incubation tend to cause premature hatches, while long storage periods, high humidity, and/or low incubation temperatures all tend to result in late hatches.

Fertility

It is unusual for all eggs in a large setting to be fertile. The average fertility for most breeds is between 80 and 95 percent during the peak of the breeding season, but early and late in the season, fertility can be significantly lower. See the chart on Addressing Incubation Problems, page 182, for common causes of poor fertility.

Hatchability

The hatchability of artificially incubated duck eggs often is 5 to 10 percent lower than that of chicken eggs. To maintain good hatchability throughout a long hatching season, in most situations it is necessary to make slight modifications to the incubation environment to adapt to the changing requirements of eggs laid early vs. late in the season.

Good setting ducks frequently hatch every fertile egg they incubate. However, under artificial incubation, the average hatchability falls between 65 and 85 percent of all eggs set, or 75 to 95 percent of the fertile eggs. See the chart on page 182 for common causes of poor hatchability.

Over the years I have had people tell me that the eggs from certain breeds or strains hatch poorly. Normally, I have found the problem was misidentified as faulty eggs, when, in fact, it was faulty incubation for those particular eggs. People frequently assume that the artificial incubation requirements are the same for all duck eggs, but they are not. If we have trouble hatching a particular kind of duck egg in our incubators, I set a number of egg clutches under ducks that have proven to be reliable setters. If the ducks hatch a normal number of these eggs, then I know the fault is not in the eggs and I need to figure out what requirements they need to hatch well in an incubator.

Natural Incubation

When you wish to hatch a moderate number of ducklings, natural incubation is often the most practical. A good setting hen is a master at supplying the precise temperature and instinctively knows just how often eggs need to be turned. She may also serve as a ready-made brooder, eliminating the need to supply an artificial source of heat.

Choosing Natural Mothers

In the Breed Profiles chart in chapter 3 (pages 28–29), the column on Mothering Ability indicates the average capability of the various breeds as natural mothers. Large duck eggs have also been hatched by turkeys, chickens, and small geese. Any chicken that is a faithful "broody" can work. Some of the chicken breeds that are more commonly used include Silkie, Cochin, and common barnyard bantams and large Old English Games, Orpingtons, and Cochins.

Clutch Size

Duck hens normally cover 8 to 12 (Muscovies 10 to 16) of their own eggs. Some hens lay such large clutches that they cannot incubate the eggs properly. In this situation the oldest eggs — those that are the dirtiest — should be removed, leaving only the number that the hen can cover comfortably. Eggs must be positioned in a single layer to hatch well, never stacked on top of one another. If too many eggs are in a nest, the result will be a poor hatch or a complete loss.

Care of the Broody Hen

Setting hens are temperamental and should not be disturbed by people or animals. It is often advantageous to isolate the broody from the rest of the flock with a temporary partition. This precaution will keep other hens from disrupting the incubation process by attempting to lay in the broody's nest. Unlike with chickens, duck hens and their nests usually cannot be moved.

To remain healthy during her long vigil on the nest, the hen must eat a balanced diet, have clean drinking water, and be protected from the hot sun. Feed and water containers should be placed several feet from the nest so that the hen must get off to eat and drink. A leave of absence from the nest for 5 to 30 minutes once or twice daily is essential to the hen's good health and will not harm the eggs.

When chicken or turkey hens are used to hatch duck eggs, they should be treated for lice and mites several days before their setting chores commence.

Multiple Broods

Muscovies (and occasionally ducks of other breeds) will frequently bring off two broods a year in northern regions or up to three or four in mild and tropical climates. Multiple broods can be encouraged by feeding hens a breeder feed in the late winter and by removing the ducklings at one to four weeks of age.

Artificial Incubation

There are many circumstances when an incubator is useful. Unlike setting hens, incubators can be used any season of the year and come in such a wide range of sizes that any number of eggs, from one to hundreds of thousands, can be set simultaneously or on alternate dates.

On the other hand, incubators must be attended to periodically each day to check the temperature and humidity, and eggs must be turned if this function is not performed automatically. The hatchability of eggs is often somewhat lower under artificial incubation than with the natural method. Also, electric incubators are at the mercy of power failures unless a backup generator is available.

Types of Incubators

Incubators are made and sold in a wide range of sizes and shapes, with varying degrees of automation. However, they can be divided into two basic types: the still-air (gravity flow) and the forced-air.

Still-Air. These models, available with electric or oil heat, closely approximate natural conditions by placing the heat source above the single layer of eggs, causing the upper surface of the eggs to be warmer than the lower portion. Still-air incubators are simple to operate, are dependable, and have low maintenance requirements but are manufactured with relatively small capacities. We have tested four models of still-air machines and have had good results with each.

Forced-Air. These incubators are equipped with fans or beaters, which move warmed air to all surfaces of the eggs and normally have multiple layers of egg trays. Forced-air machines are available with capacities of 12 to many thousands of eggs. When compared with still-air incubators, they are better suited to automatic turning of eggs and take up less floor space for larger quantities of eggs. They are also more complicated, require greater maintenance, and sell for higher prices.

Homemade. With a little ingenuity and a lot of persevering care, satisfactory hatches can be obtained in a homemade incubator consisting of a cardboard or wooden box and lightbulbs for heat (great care should be taken to avoid accidental fires). More elaborate incubators, complete with heating elements and thermostats, can also be crafted in the home shop (county Extension agents or 4-H officers often have plans available for building small incubators). In emergency situations — such as a hen's deserting her nest — an electric frying pan or heating pad can be used to finish hatching eggs. Emergency measures such as these take great diligence and patience in regulating the temperature, but they are sometimes worth the effort.

Where to Place the Incubator

Incubators of all sizes, but especially small ones, perform best in rooms or buildings where the temperature does not fluctuate more than 2 to 5°F (1.2 to 2.8°C) over a 24-hour period. Consistent temperatures are especially important for still-air incubators, which should be located in a room with an average temperature of 75 to 80°F (24 to 27°C). Do not position your machine where it will be in direct sunlight or near a window, heater, or air conditioner.

Leveling the Incubator

Incubators, especially still-air models, must be level to perform well. If the incubator is operated while askew, the temperature of the eggs will vary in different areas of the machine, causing eggs to hatch poorly and over an extended period of time.

Multistage and Single-Stage Incubation

Multistage incubation is the term used to describe the procedure of setting eggs at different times in the same incubator. Virtually all small incubators are intended to be used as multistage machines if they are filled nearly to capacity.

Typically, in multistage incubators eggs would be set once a week. Most manufacturers of these incubators specify that it is imperative that no more than one-third to one-half of the egg capacity of the machine be set in any given week. The reason is that as embryos mature they give off more and more heat. If this type of incubator is overly full of late-stage embryos, the temperature rises to dangerous levels and can kill the embryos. In multistage incubation the operating parameters (such as temperature and humidity) are maintained at the same level throughout the incubation process, in an attempt to meet the average needs of embryos at different stages of development.

Single-stage incubation refers to the procedure of filling an incubator to capacity in one setting. This method requires incubators that are equipped with coolers, which are critical for keeping the incubators from overheating during the latter stages of single-stage incubation. A big advantage of single-stage incubation is that you can provide for the specific changing needs of the growing embryos throughout the incubation period without risk of overheating them.

Operating Specifications

Manufacturers of incubators include a manual of operating instructions with their machines. This guide should be carefully read and followed. The operating instructions often cannot be adapted from one machine to another with good results, particularly if one is a still-air model and the other a forced-air. If you acquire a used incubator that does not have an instruction booklet, manufacturers are usually willing to send a new manual if you send them a request with the model number of your machine.

OUR INCUBATION PROCEDURES

During the 50 years I have been raising ducks, the following procedures have given excellent results. We have experimented with many alternatives, but these methods work best for us. The incubators we use are multi-stage, forced-air models that are rated to hold 2,500 chicken eggs. The incubator room is kept at 75 to 80°F (24 to 27°C). Our hatcher is located in a separate room adjacent to the incubator room, which greatly reduces the amount of down and dust in the incubator room — a great benefit to both the incubating eggs and the lungs of the people working there.

1. Eggs are gathered in the morning and midafternoon. They are washed the day they are laid in 100 to 110°F (38 to 43°C) water to which an antibacterial soap has been added.

2. The eggs are stored in a basement egg room at a temperature of 55 to 65°F (13 to 18°C) for 1 to 7 days.

3. Because eggs from some breeds (and even some varieties and strains within a variety) have somewhat different incubation periods, we set the eggs on a precise schedule to synchronize the hatch. Our goal is to have the ducklings ready to ship on Tuesdays, so 4 weeks prior to the desired hatch date, we start setting eggs on Monday evening in the following order: Magpie, Swedish, Saxony, Runner (White, Black, Blue, Chocolate), Aylesbury, Rouen, Pekin, Cayuga, Appleyard, Orpington, and Crested. First thing Tuesday morning we set Ancona, Harlequin, Muscovy, and other varieties of Runner eggs. Late Tuesday afternoon the Campbell and Dutch Hook Bill eggs are set. On Tuesday evening the bantam duck eggs are put in the hatchery room to start warming up and then put in the incubator on Wednesday morning.

4. Only every other tray is filled the first week. Two weeks later the other six trays are filled. This staggered setting prevents the incubator from overheating toward the end of the incubation period when the embryos are generating considerable heat.

5. From the 7th to the 24th day of incubation, the eggs are lightly sprayed once daily with 100°F (38°C) tap water.

6. From day 1 through day 24, the eggs are turned 90 degrees once an hour by an automatic turner.

7. From day 1 through day 24, the incubation dry-bulb temperature is kept at 99.2 to 99.5°F (37.3 to 37.5°C). At the beginning of the hatching season, the wet-bulb temperature for most breeds is 82 to

84°F (27.8 to 28.9°C). (For breeds or varieties whose eggs dehydrate more slowly, we start off with a wet-bulb of 78 to 80°F [25.6 to 26.7°C] in a separate incubator.) Our incubation season runs from February through June. Because eggshells become more porous as the laying season progresses, we gradually increase the wet-bulb reading over the course of the hatching season, so that by the last setting the wet-bulb temperature is 85 to 86°F (29.4 to 30°C) (plus or minus as needed).

8. On Friday evening during the fourth week of incubation (the 24th day), eggs are transferred from the incubator to the hatcher (the hatcher is not opened again until the hatch is removed). The hatcher is operated at a dry-bulb temperature of 98.5°F (36.9°C) and a wet-bulb temperature of 85°F (29.4°C).

9. On Sunday morning (the 26th day of incubation), the dry-bulb temperature is dropped to 98°F (36.7°C) and the wet-bulb temperature raised to 90°F (32.2°C). As the hatch progresses the wet-bulb temperature typically rises to 92 to 94°F (33.3 to 34.4°C) due to the damp down of all the babies that are hatching.

10. On Tuesday mornings the hatcher is opened, the trays are removed, and the ducklings are taken out. Unhatched eggs are returned to the hatcher and lightly sprayed with lukewarm tap water to soften the shell membranes, which tend to dry while out of the machine. Because every time the hatcher is opened the humidity level of the machine drops and shell membranes can dry out and become tough, we minimize the number of times the hatcher is opened.

11. After every hatch, all removable components of the hatcher are taken outside and thoroughly washed with hot water and antibacterial soap. The hatcher cabinet is vacuumed and then thoroughly scrubbed with hot water and antibacterial soap. The hatchery room is vacuumed, all surfaces are wiped down with water and soap, and the hatcher is put back together.

12. All vents of the hatcher are closed, and the inside surfaces (including the hatching trays) are sprayed with a disinfectant (I wear a respirator with charcoal filters during this procedure). Once the inside is sprayed, the hatcher door is closed and the hatcher is turned on and allowed to run for 30 minutes with the vents closed. At the end of 30 minutes, the vents are opened wide and the machine is allowed to dry out while the fan and heater are left on. The hatcher is now ready for the next hatch.

Incubation Requirements

The following is a summary of the basic incubation requirements of duck eggs. Over time you will fine-tune these procedures to meet the unique needs of your own situation and environment. Some of the factors that affect incubation requirements are climate (especially temperature, humidity, and barometric pressure), elevation, breeds being hatched, diet of the breeding stock, and the duration of the hatching season.

Setting the Eggs

Start the incubator at least 48 to 72 hours ahead of time and make all necessary adjustments of temperature, humidity, and ventilation before the eggs are set. People frequently put eggs in machines that are not properly regulated, thinking they can make fine-tuned adjustments after the eggs are in place. This practice is a serious mistake because one of the most critical periods for the developing embryo is the first few days of incubation.

Prior to being placed in the incubator, duck eggs need to be warmed up for 5 or 6 hours at a room temperature of 70 to 80°F (21 to 27°C). If cold eggs are set without this warming period, water condenses on the shells and yolks occasionally rupture.

For high-percentage hatches it is essential that eggs are incubated in the correct position. Always set them with the large end (air cell) at least slightly raised. When duck eggs are set with the air cell lowered, their chances of hatching are significantly decreased. Set only the number of eggs that fit comfortably in the tray, without crowding or stacking them on top of one another. If at all possible, do not disturb the eggs during the first 24 hours in the incubator.

Temperature

In still-air machines it is common to set the temperature at 101.5°F (38.6°C), 102°F (38.9°C), 102.5°F (39.2°C), and 103°F (39.4°C) for each consecutive week. If you do not fill the incubator with one setting, but add eggs each week, a constant temperature of 102 to 102.5°F often works satisfactorily. It is essential that the thermometer be positioned properly in still-air incubators or an incorrect temperature reading will be given. The top of the thermometer's bulb must be level with the top of the eggs. Do not lay the thermometer on top of the eggs because this practice will give a warmer temperature reading than actually exists at the level of the eggs.

We pre-warm our hatching eggs for 5 to 6 hours at a room temperature of 70 to 80°F (21 to 27°C) prior to placing them in the incubator.

Since all sides of the eggs are warmed equally in forced-air incubators, they are operated at slightly lower temperatures. For duck eggs a consistent temperature of 99.25 to 99.5°F (37.4 to 37.5°C) from day 1 through day 25 has been traditionally recommended in forced-air incubators. In single-stage incubation the trend is to start at 100 to 100.3°F (37.8 to 37.9°C) and gradually lower the temperature so that at 14 days the temperature is 99.25 to 99.5°F and by day 24 it is 98.5 to 98.7°F (36.9 to 37.1°C).

In incubators not equipped with coolers, the temperature must be watched closely for the last 10 days of incubation. During this period, an increase in temperature is not uncommon, and the thermostat may need to be adjusted slightly each day to keep the eggs from overheating. Lowering the temperature to 98 to 98.5°F (36.7 to 36.9°C) for the final 2 to 3 days during the hatch can be beneficial, since ducklings generate considerable internal heat as their respiratory system engages and they work to free themselves from their shells.

It is a good idea to use thermometers designed specifically for incubators, as they have greater accuracy and are easier to read than utility models.

Humidity

To have a large number of strong ducklings hatch, the correct amount of the contents of the eggs must gradually dehydrate. When dehydration is excessive, the ducklings are undersized and weak, making it more difficult for them to break out of the eggs. Conversely, inadequate moisture loss results in overly chubby ducklings that have difficulty turning inside the egg to crack the shell.

The rate at which the contents of eggs dehydrate is affected by the moisture level of the air in the incubator and by the porosity of the eggshell. Moisture in small incubators is usually supplied by water evaporation pans, whereas in large commercial machines, humidity is typically supplied by spray nozzles or evaporation panels.

The correct level of humidity varies widely, depending on factors such as elevation, breed of duck, egg size and shell quality, storage length and humidity prior to incubation, environmental temperature while the egg is in the nest prior to being gathered, and length of time the female has been laying eggs (the longer a duck lays, the more porous her eggs become). Therefore, it is impossible to give an exact recommendation. A good way to determine the correct humidity level is to start out by following the basic recommendations given in your incubator manual or those given here and make necessary adjustments as needed.

If your incubator is equipped with a wet-bulb thermometer or hygrometer (if it isn't, it's difficult to know what the humidity is), note that the correct reading on these instruments during the incubation period usually ranges between 80 and 84°F (26.7 and 28.9°C) on the wet-bulb thermometer, which is equal to a relative humidity of approximately 55 percent on the hygrometer. When eggs have been washed prior to incubation, the correct humidity level is normally between 1 and 2°F (0.5 and 1.1°C) higher on the wet-bulb thermometer or 60 percent on the hygrometer. As ducklings start to hatch during the last several days of incubation, the relative humidity typically needs to be raised to 88 to 94°F (31.1 to 34.4°C) on the wet-bulb thermometer to prevent the shell membranes from drying out excessively.

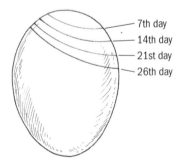

7th day
14th day
21st day
26th day

When candled, eggs that are dehydrating at the proper rate will show air-cell sizes similar to those here.

Because there are so many variables affecting the best humidity levels during the hatch, you will probably need to do some experimenting to determine what works best in your situation.

The size of the air cell (normally located at the large end of the egg) is a useful indicator of whether the contents of the eggs are dehydrating at the correct rate. The air cell's volume can be observed by candling the eggs in a darkened room. On the 7th, 14th, 21st, and 26th days of incubation, the average air-cell volumes should be approximately the same size as those in

DETERMINING CORRECT HUMIDITY LEVEL

The following procedure is how we determine the correct level of humidity:

1. Every morning at approximately the same time, we read and record the dry-bulb and wet-bulb temperatures for each incubator (be sure to take these readings before the machines are opened).

2. As the eggs are transferred from the incubators to the hatcher, we candle them and record the average air-cell size. (Air-cell size is a useful indicator for knowing whether more or less dehydration is needed.)

3. At the end of the incubation period for each setting of eggs, we add up the dry-bulb and wet-bulb readings (separately) and divide by the number of days to get averages for both dry-bulb and wet-bulb readings.

4. If the hatch is satisfactory, the machine is operated at or near those average dry-bulb and wet-bulb readings for the next batch of eggs (taking into account that as the laying season progresses shells tend to become more porous and therefore that humidity levels often need to be adjusted slightly upward).

5. If the hatch was unsatisfactory, we make adjustments accordingly.

By keeping these records on hand, we have learned over time the intricacies of each machine for different seasons of the year and for different kinds of eggs.

the illustration on page 174 (keeping in mind that the eggs of some breeds, strains, or individual females may show significant difference in air-cell size and still hatch satisfactorily). If the air cells are too large, increase the moisture in the incubator and/or decrease the amount of ventilation, being careful not to reduce the airflow so severely as to suffocate the embryos. If the air cells are too small, decrease the moisture level and/or increase the ventilation.

Carbon Dioxide

Conventional wisdom has long held that good air quality inside the incubator is critical for embryonic health. Incubator manufacturers often extolled the superior ventilation systems of their machines, with the implication that the greater the ingress of fresh air, the better — as long as appropriate humidity could be maintained.

For some time now there has been ongoing research that indicates that the carbon dioxide levels that are beneficial to developing embryos are higher than that of the air we normally breathe (which typically is about 600 ppm). The problem is that there has been little agreement about what exactly these levels should be. As little as a few months ago, an incubator salesman told me that their machines were equipped to monitor CO_2 levels but that they didn't know what levels to recommend.

The amount of CO_2 produced by an incubating egg gradually increases as the embryo grows. The current trend among some hatcheries using single-stage incubation is to close all the vents for the first 9 days of incubation, aiming for a carbon dioxide level of up to 8,000 ppm by the 13th day. After the 13th day the CO_2 level is dropped to 4,000 ppm for 3 days and then to 3,000 ppm until the 24th day. On the 24th day it is raised to approximately 4,000 ppm during the hatch.

The reported benefits for these elevated levels of CO_2 include not only higher percentage hatches but plumper hatchlings with greater vitality, and a more synchronized hatch. Obviously, excessive levels of CO_2 at any given stage of development can be deadly and must be guarded against.

Ventilation

Incubators are equipped with inlet and outlet vents that allow for air exchange. As researchers and hatchery operators learn more about the ideal environment for developing embryos, the understanding of ventilation needs has been changing. As a starting point you can open and close the vents according to the recommendations of the manufacturer of your incubator. If you want to

BURIED TREASURES

For my tenth birthday, Mom asked if there was something special I'd like to do. Without hesitation I told her I wanted to go to the city park at Waverly Lake to feed the ducks. Mom knew that there was no use in pointing out that we had a pasture full of ducks that I fed every day.

On a beautiful Sunday afternoon, my family joined in feeding stale bread and grain to my eager web-footed friends. When the larder ran out, we explored the shoreline. The previous week had brought torrential rains, and the lake was the fullest in memory. As we walked the shore, I spotted a nest of eggs lying a foot below the murky water's surface. They must be rescued! Dad voted for leaving them, pointing out they had probably been submerged for nearly a week, but the 13 pale green eggs were soon nestled in an old towel.

At home I placed the precious eggs in my incubator. The days passed slowly. On the 27th day, I heard peeping. Lifting the lid, I saw eight black-and-yellow "chipmunk"-marked Mallards. The smallest one was christened Tiny Tina and quickly became my favorite duck.

experiment with lower levels of ventilation, the information in the previous section can be used as a guideline. To know what the CO_2 levels are in your incubator, you will need to use a CO_2 tester.

Typically, in incubators that contain eggs at different stages of development (multistage incubation), the CO_2 should be about 3,000 ppm. When determining ventilation needs, always keep in mind that as the embryos develop, they are producing greater quantities of CO_2. The amount of ventilation required is relatively small but nevertheless essential to the well-being of the developing ducklings. Ventilation demands increase at hatching time.

Turning during Incubation

Incubating eggs must be turned for successful hatches. While some eggs will hatch if turned just once every 24 hours, for consistently high-percentage hatches, they should be turned a minimum of two to three times daily at approximately 8- to 12-hour intervals. Turning is most critical during the first 2 weeks of incubation.

If your incubator is equipped with automatic turners, you may see additional benefit from turning the eggs more frequently. Some commercial incubators are programmed to turn eggs as frequently as every 15 minutes. However, for the small producer, turning once every 1 to 2 hours can give excellent results. Automatic turners can be started as soon as the eggs are set.

If your incubator is not equipped with an automatic turner and you need to open it in order to turn the eggs, it is best to start turning 12 to 24 hours after the eggs are set (and the temperature has stabilized), and to turn them only three times a day in order to minimize opening the incubator. For best results eggs need to be rotated at fairly regular hours and revolved at least one-third of the way around at each turning. Some studies indicate that turning eggs a full 180° improves hatchability. Eggs must be turned gently to avoid injury to the embryo. Turning can be discontinued 3 days before the hatch date in still-air incubators; up to 7 days in forced-air machines.

When eggs are turned manually, it is helpful to mark them with an X and an O on opposite sides with a wax or graphite pencil. Liquid inks — such as those in felt-tipped pens — should not be used, since they clog shell pores and can poison the embryo. After each turning all eggs should have the same mark facing up.

Cooling

For best results when using still-air incubators, eggs can be cooled once daily — except during the 1st week and the last 3 to 4 days of incubation (if eggs are sprayed as outlined below, additional cooling gives minimal benefits in forced-air machines). When the room temperature is 70 to 75°F (21 to 24°C), the top of the incubator should be removed or the trays taken from the incubator and the eggs cooled for 5 to 8 minutes a day the 2nd week, 8 to 10 minutes daily the 3rd week, and 12 to 15 minutes the first 3 to 4 days of the 4th week. The length of the cooling depends on the size of the eggs: for smaller eggs use the shorter cooling times.

To protect against your becoming sidetracked and forgetting the eggs as they cool, it is wise to set a timer for the appropriate number of minutes. While eggs normally hatch if left without heat for several hours once or twice during incubation (except during the 1st week and the last 5 days, when low temperatures can be disastrous), repeated overcooling will retard growth and can be fatal.

If you ever find that the temperature in the incubator is excessively high—by more than 1.5°F (0.8°C)—immediately cool the eggs for at least 10 minutes and make adjustments to correct the problem.

Spraying or Sprinkling?

After considerable experimentation we have found that we get consistently higher-percentage hatches by lightly spraying duck eggs once a day with lukewarm distilled or well water from the 6th to the 24th day of incubation. It may seem counterintuitive, but we have noticed that eggs that are sprayed regularly dehydrate more than unsprayed eggs.

To prevent the egg membranes from drying out and becoming tough during the hatch, it is sometimes helpful to lightly spray or sprinkle duck eggs with warm water 48 and again 24 hours before the calculated hatching time. However, we have found that if the eggs dehydrate the proper amount throughout the incubation period, spraying during the hatch is unnecessary.

Candling

The best time to candle duck eggs to check fertility is on the 5th to 7th days of incubation. If eggs are candled prematurely, it is more likely that fertile eggs will be missed and accidentally discarded.

Eggs are candled in a darkened room with an egg candler or flashlight. On the 7th day fertile eggs reveal a small dark spot with a network of blood vessels branching out from it, closely resembling a spider in the center of its web. Infertile eggs are clear, with the yolk appearing as a floating shadow when the egg is moved from side to side.

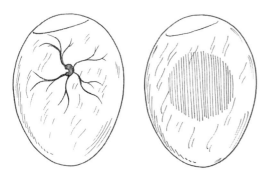

On the 7th day a fertile egg (left) has a spidery network of blood vessels, while an infertile egg (right) is clear.

A blood ring visible when candling indicates that the embryo has died.

Sometimes embryos begin to develop but then perish within several days. When this happens, a streak or circle of blood is visible in an otherwise clear egg. Contaminated and rotten eggs often exhibit black spots on the inside of the shell, with darkened, cloudy areas floating in the egg's interior. All eggs not containing live embryos should be removed from the incubator.

Contaminated and rotting eggs give off harmful gases and may ooze or explode, covering the other eggs and the incubator's interior with putrid-smelling, bacteria-laden goo that is difficult to clean up. To reduce the chances of blowouts, we candle our duck eggs on the 7th, 12th, 19th, and 24th days of incubation.

The Hatch

All the time and effort invested in producing and incubating eggs is rewarded by the hatch. Those first muffled chirps of the ducklings are sweet music.

Normally, the first eggs will be pipped (have a small crack in the shell) 48 hours prior to the hatch date. Ducklings require 24 to 48 hours to completely rim the shell and exit. Newly hatched birds are wet and exhausted and typically are left in the hatcher for 12 to 24 hours to gain strength and dry off.

If your hatcher has adjustable air vents, they may need to be regulated to give hatchlings additional air at the peak of the hatch. However, do not open them so wide that the humidity level drops. During the hatch you may find it necessary to place extra water-evaporation pans or pads in small machines to maintain an adequate level of humidity with the increased air circulation.

SEPARATE HATCHER HIGHLY RECOMMENDED

Rather than incubating and hatching eggs in the same machine, there are significant advantages to incubating eggs in an incubator for the first 23 to 24 days and then transferring them to a separate machine for the last few days during the hatch. By using a separate hatcher, the unique turning, temperature, and humidity requirements of eggs at different stages can be better met, the incubator is kept cleaner, and the hatcher can be thoroughly cleaned and disinfected after each hatch.

Evaporation pans should be covered with screen or hardware cloth to make certain that the ducklings cannot drown.

After most of the ducklings have hatched, the relative humidity can be lowered to 50 percent (82 to 84°F [27.8 to 28.9°C] on the wet-bulb thermometer) for a couple of hours so they will fluff out properly.

Removing Ducklings from the Incubator

When the ducklings are dry, remove them from the incubator. Before opening the machine, prepare a clean container — with sides at least 6 inches (15 cm) high — with bedding that gives good footing, such as shredded wood excelsior pads, mold-free straw or hay, or old towels (make sure the towels don't have strings that the hatchlings can swallow or get wrapped around their tongues or legs). While transferring ducklings, work quickly and gently, discarding empty shells and pipped eggs containing birds that are obviously dead. It is best if the room temperature is at least 70°F (21°C). Any wet ducklings can be left in the incubator until dry.

Help-Outs

Occasionally at the end of a hatch, a few live ducklings may still be imprisoned within partially opened shells. Assistance can be given by carefully breaking away the shell just enough (stop if bleeding occurs) so that the hatchling will be able to exit on its own. Keep in mind that the closer the incubation environment was to the ideal throughout the incubation period for the particular eggs being hatched, the less likely that hatchlings will need assistance.

Some birds that are assisted from the shell develop into fine specimens. However, if the reason they were unable to hatch without assistance is due to an inherited weakness or deformity, they should not be used for breeding purposes at maturity. When it is understood that the hatch is a fitness test given by nature to help cull out the weak and deformed, we can take a more realistic view of helping ducklings from the shell. Some breeders permanently mark (by notching or perforating the webbing between the duckling's toes with a chick toe punch available from poultry supply dealers) any ducklings they assist in hatching so they are not used for breeding purpses unless they have exceptional qualities as mature birds.

ADDRESSING INCUBATION PROBLEMS

Symptoms	Common Causes	Remedies
More than 10 or 15% clear eggs when candled on 7th day of incubation	Too few or too many drakes	Correct drake-to-hen ratio
	Old, crippled, or fat breeders	Young, active, semifat breeders
	Immature breeders	Use breeders seven months or older
	Breeders frequently disturbed	Work calmly around breeders
	No swimming water for mating	Swimming water for large breeds
	First eggs of the season	Don't set eggs laid 1st week
	Late-season eggs	Don't set eggs when males molt
	Medication in feed or water	Avoid medicating breeders
	Eggs stored more than 14 days	Set fresher eggs
Blood rings on 7th to 10th day	Faulty storage of eggs	Proper storage prior to setting
	Irregular incubation temperature	Adjust machine ahead of time
Ruptured air cell	Rough handling or a deformity	Handle and turn eggs gently
Yolk stuck to shell interior	Old eggs that haven't been turned regularly during storage	Turn eggs daily if they are held for more than 5 days
Dark blotches on shell interior	Dirt or bacteria on shells causing contamination of the inner egg	Wash eggs with water and disinfectant soon after gathering
More than 5% dead embryos between 7th and 25th day of incubation	Inadequate breeder diet	Supply balanced diet
	Highly inbred breeding stock	Introduce new birds to flock
	Incorrect incubation temperature	Check accuracy and position of thermometer
	Periods of low or high temperature	Check temperature often; don't overcool eggs
	Faulty turning during incubation	Turn at least three times daily
Early hatches	High incubation temperature	Lower incubation temperature
	Low incubation humidity	Adjust humidity

Symptoms	Common Causes	Remedies
Late hatches	Low incubation temperature	Raise incubation temperature
	High incubation humidity	Adjust humidity
Eggs pip but do not hatch; many fully developed ducklings dead in shells that aren't pipped	High humidity during incubation	Decrease humidity
	Low humidity during incubation	Increase humidity
	Eggs chilled or overheated during last 5 days of incubation	Protect eggs from temperature extremes during this period
	Low humidity during the hatch causing egg membrane to dry out	Last 3 days, raise humidity to 75% and sprinkle eggs daily
	Poor ventilation during hatch	Increase air flow during hatch
	Disturbances during hatch	Leave incubator closed
Eggs pipped in small end	Eggs incubated in wrong position	Position eggs with small end lower than large end
	Chance	No remedy
Sticky ducklings	Probably low humidity during incubation and/or hatch	Increase amount of moisture
Large, protruding navels	High temperature	Lower temperature; check thermometer
	Excessive dehydration of eggs	Increase humidity level
	Bacterial infection	Improve sanitation practices
Dead ducklings in incubator	Suffocation, overheating, or high bacteria count	Increase ventilation during the hatch, watch temperature carefully, or improve sanitation
Spraddled legs	Smooth incubator trays	Cover trays with hardware cloth or other nonslip surface
More than 5% cripples other than spraddled legs	Inadequate turning during incubation	Turn eggs a minimum of three times daily
	Prolonged periods of cooling	Don't forget eggs when cooling
	Inherited defects	Select breeders free of defects

INCUBATION CHECKLIST

❑ Provide adequate nests furnished with clean nesting material.

❑ Gather eggs in the morning and again in the afternoon to protect them from temperature extremes, breakage, soiling, and predators.

❑ Handle eggs with clean hands or gloves, and avoid shaking, jolting, or rolling them.

❑ Store eggs in clean containers in a cool, humid location, and turn them daily if held longer than 5 days.

❑ Set only clean (or reasonably clean) eggs with strong shells.

❑ Start the incubator well in advance of using. Make adjustments to temperature, humidity, and ventilation before eggs are set.

❑ When filling the incubator, position eggs with their air cells at least slightly raised, and do not overcrowd eggs in the tray.

❑ If at all possible, do not disturb eggs during the first 24 hours in the incubator.

❑ Gently turn eggs at least three times daily at regular intervals from the 2nd to the 25th day (2nd to 32nd for Muscovies).

❑ Make sure the thermometers used in your incubator are accurate, to ensure that you are using appropriate temperatures.

❑ Keep the humidity in your incubator at a level that allows for the appropriate dehydration of the eggs. Remember, different kinds of eggs require different levels of humidity.

❑ When using a still-air machine, cooling eggs can improve hatches.

❑ Lower the incubator temperature 1 to 1.5°F (0.5 to 0.8°C) for the hatch.

❑ Increase the relative humidity while eggs are hatching.

❑ Do not open the incubator except when absolutely necessary during the hatch.

❑ Leave ducklings in the machine until they are dried.

❑ Clean and disinfect the incubator after each hatch.

Incubator Sanitation

While the incubator supplies the correct conditions for the embryo to develop, it also provides an excellent environment for the rapid growth of molds and bacteria. Consequently, it is essential that the incubator and the hatcher be kept as clean as possible.

At the conclusion of each hatch, clean and disinfect the incubator and hatcher. If the eggs are set to hatch at various times, the water pans should be emptied, disinfected, and returned with clean, lukewarm water, and the duckling fuzz removed from the incubator or hatcher with a damp cloth or vacuum cleaner.

At the end of the hatching season, thoroughly clean the machines and store in a dry, sanitary location.

13

Rearing Ducklings

THE DOWNY YOUNG OF ALL TYPES OF POULTRY are charming, and day-old ducklings are no exception, with their bright eyes, tiny wings, and miniature webbed feet. Aided by shovel-like bills, they are soon eating an amazing quantity of food, all the while growing at an equally astounding rate. In only 8 to 12 weeks, a newly hatched duckling is transformed into a young adult.

Basic Guidelines

Of all domestic fowl, in many circumstances the young of ducks are the easiest to raise. It is common for every duckling in a brood to be reared to adulthood without a single mortality. They can withstand less than optimum conditions well — although that is no reason to mistreat or neglect them. If you put into practice the following management guidelines, raising duck-lings can be a pleasant and trouble-free task.

1. Keep them warm and dry, and protect them from drafts. Ducklings that are cold lose their appetites, and when wet they chill rapidly and will die if not dried and warmed promptly. Drinking-water containers should be designed so that ducklings cannot enter them and become excessively wet.

2. Maintain them on dry bedding that provides good footing or on wire or plastic flooring (that they cannot get their legs stuck in). Smooth, slick floors are the leading cause of spraddled legs. Many internal parasites, molds, and disease organisms thrive in damp and filthy bedding.

3. Supply fresh feed that provides a balanced diet, and make any dietary changes gradually to avoid possibly lethal digestive-tract enteritis. A proper diet that is formulated to meet all of the dietary requirements of ducklings is essential for disease resistance and normal growth. Feed that is moldy, deficient in nutrients, or contains certain additives can cause stunted growth, sickness, or death. Never use feed containing arsenicals or the drug roxarsone. (If nonmedicated feeds are unavailable in your locale, rations containing amprolium, chlortetracycline, neomycin, oxytetracycline, sulfaquinoxaline, or zinc bacitracin have been used successfully by many duck growers and should work satisfactorily for you if one of these drugs is incorporated in the feed at the proper dosage. See Poisoning from Medication, chapter 17, page 296.)

4. Provide a constant supply of fresh water. Ducklings suffer a great deal when drinking water is not available, particularly right after they have eaten. (Never add Ren-O-Sal tablets to the drinking water of ducklings — the active ingredient roxarsone can be deadly to young waterfowl.)

5. Furnish adequate floor space and fresh air. Forcing ducklings to live in crowded conditions is one of the leading causes of feather eating, wing disorders, and disease.

6. Protect them from predators. Tame and wild predatory animals and birds find unprotected ducklings easy prey. The clumsy feet of humans and large animals also snuff out the lives of many young birds.

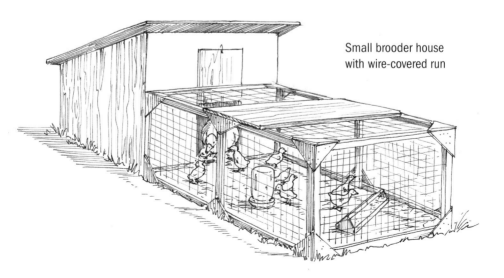

Small brooder house with wire-covered run

Natural Brooding

For the home poultry flock, natural brooding can have advantages. It eliminates purchasing or constructing special brooding equipment and supplying artificial heat. When allowed unlimited range in an area with abundant natural duck foods, hen-mothered ducklings learn to forage for their food at a young age and will be partially self-sufficient if there is a plentiful supply of insects, earthworms, slugs, snails, tender and succulent greens, and wild seeds. Ducklings have been successfully brooded by duck, turkey, bantam, or large chicken hens.

A day-old Mallard exhibiting the bold black and yellow markings that are typical of all Gray varieties of ducklings.

By two weeks of age, the down color has faded and feathers can be felt protruding from her tail and sides.

At four weeks, feathers cover the face, shoulders, underbody, and tail and have begun to appear on her back.

Now eight weeks old, the once soft, cuddly duckling is fully feathered and nearly the size of her parents.

Managing the Hen and Her Brood

The best management practice in dealing with a hen and her brood of duck-lings is to bother them as little as possible. Your main concern is to protect them from predators and to keep the ducklings from becoming soaked during their first several weeks of life outside the shell.

Free-roaming turkeys and chickens often leave ducklings stranded, so it is a good idea to enclose foster hens with their broods in a dry building or pen for the first 2 to 4 weeks to protect them from rain, cold winds, predators, and rodents. Concentrated feed can be supplied to get them off to a fast start.

The same procedures are recommended for duck hens and their broods, although it is normally safe to give them freedom in a large pen not occupied by other fowl after the ducklings are several days old. Duck hens are not as likely to jump or fly over barriers such as fences and leave the little ones behind.

Confine hens and their broods in a secure pen or building each night until the ducklings are six to eight weeks old. This significantly reduces the possibility of losses to predators.

Hens can be aggressive toward unknown ducklings and may injure or kill them. It is usually best not to confine two or more hens together with their broods, particularly during the first week or two.

Allowing a new brood to mix with an established flock can also be dangerous for the young ones, especially when ducks are penned in close confinement and the strange ducklings may threaten the territorial boundaries of the adults. If your flock of ducks is permitted to roam freely, ducklings may not be bothered. In any case, newly hatched ducklings should be allowed with adult birds only if you can be on hand to remove them if a problem develops.

Most children love to hold baby ducks. To prevent injury from dropping or suffocation, ducklings should be held on the palm of a cupped hand while the second hand is placed over the top, as demonstrated here by our nephew, Christopher, shown at seven years of age.

Giving Hens Foster Ducklings

Sometimes it is desirable to give hens foster ducklings to brood. If a hen already has young, the foster ducklings should be approximately the same size and color as her own. Hens who set but for one reason or another do not hatch any ducklings — or chicks, in the case of a chicken hen — often will accept a foster brood. Giving ducklings to a chicken hen that has chicks of her own usually does not work out, but it can be attempted in an emergency situation. To lessen the possibility of rejection, sometimes it helps to slip foster ducklings under their new mother after dark.

Novice duck raisers sometimes lock a hen (that has been setting for a short time or not at all) in a pen with ducklings and expect her to mother them. Almost without exception these attempts fail and frequently result in ducklings being brutally attacked. Unless an exceptional hen is discovered, it is not wise to give foster ducklings to hens that have not set their full term.

The number of ducklings a hen can brood depends on her size, the size of the ducklings, and the weather conditions. Typically, during relatively mild weather, a duck hen can successfully brood 8 to 12 of her own young. Bantam hens can handle 4 to 8 ducklings, large chickens 12 to 18, and turkey hens up to 25. The most important factor is that all the ducklings can be hovered (covered) and warmed by the foster mother at the same time.

Artificial Brooding

If a broody hen is not available or you are raising a large number of birds, it will be necessary to brood ducklings artificially. When sound management practices are used, this method produces good results.

The Brooder

Brooders provide warmth for the ducklings from the time they hatch until they no longer need supplemental heat. Brooding equipment can be purchased from stores handling poultry supplies or can be fabricated at home.

Homemade Brooder

A few ducklings can be brooded with a lightbulb that is positioned in a box or wire cage. Research indicates that blue bulbs are best because they reduce the incidence of feather eating and are gentler on the ducklings' eyes.

The wattage required depends on the size of the box and the room temperature. I prefer to use several 40-watt bulbs rather than a single larger one since this provides some protection if one burns out while you are away. The box or cage must be big enough to allow ducklings to move away from the heat when they desire and to provide adequate room for the waterers to be located far enough away to prevent water from being splashed on the bulb — which can cause hot bulbs to shatter.

When bulbs of over 40 watts are used, they must be located out of the reach of ducklings to prevent burn injuries. To prevent fires or the asphyxiation of ducklings from any smoke produced by smoldering materials, even bulbs of low wattage must not touch or be close to flammable substances.

CAUTION

Every year we hear reports of fires caused by brooding accidents. **Make doubly sure** you securely fasten all light fixtures to prevent them from accidentally falling or touching any flammable materials.

A light reflector and clamp (a.k.a. a clip light) make an inexpensive "hover brooder." They are available at hardware stores and, when not used for brooding, come in handy around the shop and house. The reflector can be clamped (and for safety should be additionally secured with wire) to the side of a cage or box and will brood up to a dozen ducklings if the ambient temperature of the room is 70°F (21°C) or warmer. As the young ones grow, the heat source can be raised to an appropriate height.

A variety of homemade brooders can also be crafted. Hover-type brooders can be built with a plywood or sheet-metal canopy and porcelain light fixtures. Better still is the washtub brooder built and used by a retired coal miner friend of ours. He outfitted an old zinc washtub with several light fixtures and three adjustable legs. With just a little work and a lot of ingenuity, he created a safe brooder that gave dependable service for many years. Large plastic storage containers hold up better than cardboard boxes for brooding small numbers of ducklings due to the playful, wet habits of these waterfowl. Use your imagination!

Heat Lamp

A simple method for brooding ducklings is with 250-watt, infrared heat lamps. Suspended 18 to 24 inches (45 to 60 cm) above the litter, each lamp will provide adequate heat for 20 to 40 ducklings, depending upon the size of the birds and the ambient temperature. We always use at least two lamps in case one fails.

When heat lamps are used, extreme care must be taken to prevent fires. These lamps must never be hung with the bottom of the bulb closer than 18 inches from the litter or so that any part of the bulb is near flammable material such as wood or cardboard. Always make sure heat lamps or other types of heaters are securely fastened to something that is sturdy and will not fall over or collapse.

Caution: Always use porcelain light fixtures with heat lamps — never plastic! Be sure heat lamps are positioned so the bulb is a minimum of 18 inches (45 cm) away from the bedding or sides of a combustible brooding frame.

Battery Brooder

Manufactured battery brooders traditionally were all metal, although some companies are now selling them with plastic components. Some folks make their own brooders using plywood sides. A battery brooder is equipped with a wire floor, removable or slide-out dropping pan, thermostat, electric heating element (in homemade ones, some people use lightbulbs), and feed and water troughs. They can be purchased and used in individual units or stacked on top of one another. Battery brooders are commonly used for starting chicks but can also be used for brooding ducklings up to an age of two to six weeks, depending on the breed size. While new units are expensive, this type of brooder requires limited floor space, is relatively easy to clean, protects young birds from predators, and can provide sanitary conditions. An added bonus is that with their wire-covered floors newly hatched ducklings will not consume bedding as they learn what is food and what is not. We start all our hatchlings in this type of brooder for the first 4 to 10 days.

A battery brooder provides heat, troughs for food and water, and good protection from predators.

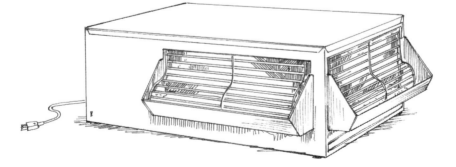

Hover Brooder

Available with gas, oil, or electric heat, this brooder consists of a thermostatically controlled heater that is covered with a canopy. The brooder is supported with adjustable legs or suspended from the ceiling with a rope or chain and is easily set to the proper height as the birds grow. Water fountains and feeders are placed away from the canopy where the ducklings drink, eat, and exercise in cooler temperatures. Hover brooders are available in sizes that are rated at 100- to 1,000-chick capacity. The number of ducklings under each unit should be 50 to 100 percent of the given capacity for chicks, depending on the duck breed being raised. For the largest and fastest-growing meat breeds, 50 to 60 percent of chick capacity often works best.

When hung from a chain, a hover brooder — shown here with draft guard — can be raised as ducklings grow in size.

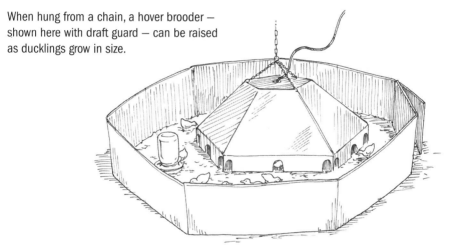

Draft Guard

The use of a draft guard around hover brooders and heat lamps is recommended for the first week or two. The guard protects ducklings from harmful drafts and prevents them from wandering too far from the heat and piling up in corners. Commercially produced guards constructed of corrugated cardboard can be purchased and used several times. Homemade guards can be fashioned with 12-inch boards or welded wire that is covered with burlap or paper feed sacks. (Whenever fabric feedbags or burlap is used, care needs to be taken that the birds do not eat the strings or strands of the bags, which can get wrapped around their tongues or cause digestive impactions.) As a general rule, the guard should form a circle 2 or 3 feet (0.6 or 0.9 m) from the outside edge of the hover brooder canopy.

Heat

Ducklings require less heat than chicks do. The exact amount of heat needed depends on factors such as humidity, air movement, ambient air temperature, and breed of duckling. More heat should be provided under humid or drafty conditions or cold outside temperatures; for tiny ducklings from small breeds; or when ducklings have been stressed in some manner.

OUR BROODING AND REARING PROCEDURES

There are a variety of ways that ducklings can be brooded and reared successfully. The procedure used will depend on factors such as the number of ducklings to be raised, the facilities available, climate, the purpose of the ducks (pets, show, breeding stock, laying stock, broilers for meat, or roasting birds), and personal preferences.

Our main purpose in raising ducks is to help preserve and distribute heritage breeds, varieties, and strains. Therefore, we have devised a system that produces ducks that are healthy, physically vigorous, productive, long-lived, and safe from predators. We also work hard to provide an environment in which our ducks can enjoy themselves and be ducks! As early as it becomes safe, they spend their days outside in grassy yards or pastures where they can enjoy natural foods, fresh air, sunshine, bathing water, and plenty of exercise.

1. Approximately 12 hours after hatching, the ducklings are removed from the hatcher, placed in chick-shipping boxes, and left in the 75 to 80°F (24 to 27°C) hatchery room for 24 to 48 hours.

2. Two days after hatching, the ducklings are put in a battery-type turkey brooder (with wire floor) that has been preheated to 95°F (35°C). In addition to the water trough along the front of the brooder, we place a quart waterer inside for each 15 to 20 ducklings to ensure that they will readily find the water during the first 2 days. We also cut up tender young grass, dandelion, or leaf lettuce into pieces about ⅛-inch square and sprinkle it on the water to further entice them to drink.

3. The trough feeder is filled to the top with 18 percent protein waterfowl starter/grower crumbles. Chick-size granite grit and

The actions of the ducklings are the best guide to the correct temperature. If ducklings are noisy and huddle together under the heat source, they are cold, and additional heat should be supplied. When they stay away from the heat, or pant, they are too warm, and the temperature needs to be lowered. The proper amount of heat is being provided when ducklings sleep peacefully under the brooder or move about freely, eating and drinking.

old-fashioned oatmeal (uncooked) are lightly sprinkled on top of the crumbles once a day.

4. At five to ten days of age (depending on the weather), the ducklings are placed in a brooder building having solid floors covered with 1½ to 3 inches (4 to 7.5 cm) of dry pine, fir, or cedar shavings or coarse sawdust. Heat is provided by 250-watt heat lamps 18 inches (45 cm) above the floor. Feed is supplied in hanging feeders and water in vacuum or automatic waterers located on outside water porches; these floors are covered with plastic poultry flooring with openings ⅞ inch × ⅞ inch (2.25 × 2.25 cm) or welded wire with openings ½ inch × 1 inch (1.25 × 2.5 cm) in size.

5. Chopped lettuce, dandelion greens, or tender young grass is fed several times a day in a quantity that will be cleaned up in approximately 1 hour.

6. Starting at 10 to 14 days of age, 16 percent–protein grower pellets (that include 20 percent oats) are gradually introduced by mixing with the starter crumbles over the course of several days.

7. As soon as the weather permits, the ducklings are allowed access to grassy yards. The heat lamps are left on in the brooder building for the ducklings to use when they get cool. Adult geese are kept with them and/or around the perimeter of these yards to discourage airborne or ground predators.

8. Every night and during daytime inclement weather, the ducklings are walked into the brooder building and locked in.

Our rule of thumb is to have the brooder temperature at 95°F (35°C) the first 2 days, lowering it to 90°F (32°C) for the remainder of the first week, then lowering it 5 to 7°F (2.75 to 4°C) each successive week. After the first week, especially for ducklings of larger breeds, you can count on being able to lower the temperature 1°F (0.5°C) per day until no supplemental heat is needed to equal the low ambient temperature of the day. Once ducklings are six to eight weeks old and well feathered, they can withstand temperatures down to 50°F (10°C) or lower, but they must be protected from drastic temperature fluctuations.

Even at the start of the brooding period, it is extremely important that ducklings are able to get away from the heat source when they desire. Overheating is almost as damaging to ducklings as chilling.

Brooding without Artificial Heat

If you don't have electricity available for artificial heat, you can still brood ducklings. One technique is to keep ducklings in a box near the stove or furnace until they are large enough to be put outside. If you use this method, be careful not to place the container too close to the heat, which could overheat the ducklings or, worse yet, start a fire.

Another procedure that works satisfactorily in mild weather is to utilize the body heat given off by the ducklings to warm themselves. A well-insulated box is the basic equipment needed. The floor of the container should be covered with 2 to 4 inches (5 to 10 cm) of dry, heat-retaining bedding such as clean rags; chopped grass-hay, straw, or pine; fir or cedar shavings; or coarse sawdust. If a fine bedding such as sawdust is used, cover it with a hemmed cloth or burlap the first day or two to prevent ducklings from harming themselves by eating the particles of wood before they learn what their food is. It is important that the edges be hemmed, to prevent the ducklings from pulling on loose string ends that could become tangled around their tongues or legs or cause digestive impactions.

The top and sides of the box can be draped with a layer of old towels, blankets, or sleeping bags. Be sure that the little ones have sufficient air to prevent suffocation. Since heat rises, the ducklings will stay warmer if the "hot box" you design has just enough headroom for them to move around comfortably after the bedding is in place. Also keep in mind that the smaller the inside area is, the cozier the occupants will be. Obviously, as the ducklings grow, their space needs will increase.

A small door should be made in the side of the brooder box where the ducklings can exit to eat and drink. They soon learn to go back inside when they're chilly, although you'll probably need to give them a helping hand the first day or two.

A clever brooding system we observed in Guatemala consisted of an earthen oven located in the center of one end of a brooder building. The horizontal, thick earthen flue (whose profile was a low flattened tube) ran through the middle of the brooder building floor to the opposite side and exited into the chimney. Hatchlings could sit on top of the flue or as close as they were comfortable. The oven, which was made out of a mix of the local clay and sand, was fed from the outside.

Floor Space

When ducklings of medium- to large-size breeds are started on litter, allow a minimum of 0.75 square feet (0.7 sq m) of floor space per bird for the first two weeks, 1.75 square feet (0.15 sq m) until four weeks of age, 2.75 square feet (0.25 sq m) until six weeks, and 3 to 5 square feet (0.3 to 0.5 sq m) per bird thereafter. At three to four weeks of age, give ducklings access to an outside pen or yard during mild weather, allowing 10 square feet (1 sq m) of space per bird. This additional space will help keep the inside bedding drier and reduce sanitation problems. Bantam breeds require approximately one-quarter to one-half the floor space of large breeds. If wire or plastic poultry flooring is used, the amount of space needed per duckling can be reduced by approximately 50 percent.

If ducklings are overcrowded, the chances for feather eating and the likelihood of poor sanitation conditions increase.

Litter

Ducklings consume a huge volume of water, so it is exceedingly important that a thick layer of absorbent, mold-free litter (or plastic or wire poultry flooring) is used to keep the brooding area from becoming sloppy. Materials such as clean pine, fir, and cedar wood shavings and coarse sawdust; peanut hulls; peat moss; crushed corncobs; coffee bean hulls; flax; or straw can be used for bedding. Whatever bedding you use, make sure it is mold free and has not been contaminated with treated products. Sawdust from cabinet shops can contain harmful materials and generally should be avoided.

Begin with 3 to 6 inches (7.5 to 15 cm), and add new litter as required. Soggy, caked bedding can be removed and replaced with dry material, although a thin layer of dry bedding sprinkled on top as needed often suffices. In some situations, it works well to stir soiled bedding before sprinkling on the top coat. In warm weather flies can become a problem if the quality of the litter is allowed to deteriorate.

Yards and Pastures

Ducklings can be given access to a yard or pasture as soon as they are comfortable outside if they are provided adequate protection from predators in your area. The exercise, fresh air, and sunlight are beneficial to their health and a tremendous aid in developing strong legs and wings. Ducks are not true grazers as are geese; however, they will eat some tender grass and other palatable greens when available, which enriches their diet with health-promoting nutrients not found in grain-based feeds. Ducklings cannot digest mature, dry, or coarse grasses.

When allowed outside, ducklings should always have ready access to a dry, protected area in case of rain, until they are fully feathered out at seven to ten weeks of age. Shade should always be available.

Water

Being waterfowl, ducklings love to swim and they consume three to four times more water than chickens do. In an effort to keep water receptacles clean, people sometimes add common household bleach to the drinking and swimming water of their ducks. This can be a dangerous practice because bleach can ravage the desirable bacteria in the birds' digestive tracts, and it is damaging to their feathers.

Swimming Water

It is not necessary to have swimming water for ducklings, even though they thoroughly enjoy going for a paddle within days after hatching. Even when ducklings are brooded by a duck hen, it is safest to keep them out of water until they are a couple of weeks old.

To protect ducklings from drowning, all water containers that they can enter should have gently sloping sides with good footing to allow tired and wet swimmers to exit easily. If you supply swimming water in receptacles having steep or slick sides, an exit ramp must be provided if drowning losses are to be avoided. Drowning and becoming soaked and chilled while swimming are the leading causes of duckling mortalities in home duck flocks.

Drinking Water

To thrive, ducklings must have a constant supply of drinking water. To ensure good hydration, when we first put ducklings in the brooder they are given water with finely chopped tender greens sprinkled on the surface for approximately half an hour before they are given duckling starter. Drinking fountains should be designed so that young ducklings cannot get into the water and become soaked. Under natural conditions, ducks of all ages wash excretions from their eyes throughout the day. Therefore, to help them keep their faces clean, it is helpful to have drinking water deep enough for them to wash their eyes and bills in.

Sufficient receptacles should be provided so their contents are not quickly exhausted and the ducklings left without water. Placing water containers on screen-covered platforms is a big help in keeping the watering area dry and sanitary. Waterers should be rinsed out daily.

If maximum growth rate is not required, once ducklings are several weeks old, they can go for 8 to 10 hours at night without water if they do not have access to feed.

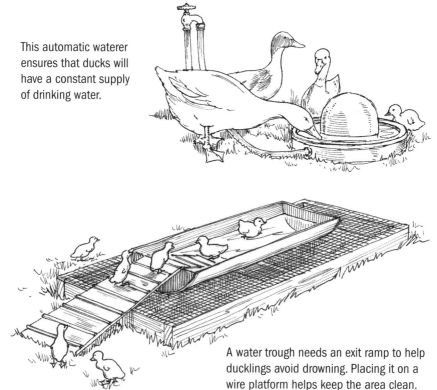

This automatic waterer ensures that ducks will have a constant supply of drinking water.

A water trough needs an exit ramp to help ducklings avoid drowning. Placing it on a wire platform helps keep the area clean.

ONE ENVIRONMENTALIST'S APPROACH TO PREDATORS

Levi Fredrikson of Watermelon Sugar Farm, located in Corvallis, Oregon, looks at dealing with predators through the lens of environmental stewardship.

Raising a small flock of ducks on ten acres of wooded pasture at the west edge of town has brought us great joy as well as many opportunities to learn. By far, predators have been the single greatest challenge in pasture-raising all manner of fowl. Losing your gentle birds to woodland creatures evokes an utterly helpless feeling. We began to view it as combat with the hawks, bobcats, owls, foxes, raccoons, weasels, opossum, coyote, skunks, and potentially even cougar. While our reaction was to trap or kill our adversaries, we realized flaws in that approach. Though trapping may lessen predator pressures temporarily, it may open a niche for another predator to fill or reduce pressures on a lesser predator (coyote and bobcat are known to eat raccoon, for example). Further, those predators were a part of the local ecosystem for many years before we began living among them and we view it as our responsibility to learn to live among them peaceably. Therefore, we amended our management to discourage local wildlife from visiting the fowl buffet. Here are our management strategies.

1. First and foremost, we lock up our birds by sundown every night (if we are out of town, we make arrangements with friends). Coops are the final line of defense and should be created with all predators in mind. Just one raccoon visit can wipe out your whole flock. Owls will gladly stop by at twilight to look for a meal. Never underestimate the tenacity and ingenuity of nocturnal hunters.

2. Five-foot-[1.5 m] high, welded-wire fencing (2" × 4" openings) partitioning the birds into one-acre [0.5 ha] pasture paddocks has greatly reduced our losses to overzealous wandering. Heavily wooded areas are a favorite of lurking bobcats and coyotes, while large raptors patrol open pasture. Two strands of electrified 17-gauge aluminum wire attached to the T-posts on the outside of the welded wire give us peace of mind. One wire is mounted on 6-inch-long insulators 4 to 6 inches [10 to

15 cm] above the ground to keep out diggers. The second is mounted on insulators 5 inches [12.5 cm] above the top of the welded wire fence to stop climbers such as raccoons. We advocate learning about proper electric fence construction from an experienced individual prior to putting it into use.

3. We strive to follow an ingenious model used in West African villages. In the villages, concentric circles extend outward from the homestead and livestock are placed according to susceptibility to predator attack (i.e., smaller animals remain within the inner circles near the home). On our farm, young ducklings are adjacent to our home in a small pen with plenty of cover, juveniles are gradually introduced to larger pasture with their mother as a guide, and adults are staggered outward according to their size (with turkeys in the outer paddocks).

4. We realized it is best to use livestock guardian dogs and llamas as enhancement to, not as a replacement for our existing predator protection. Many of our fowl-raising friends have had great success with dogs and reported some success with llamas, but they can't be your sole defense. Be sure guardian dogs are diligently trained and don't pose any threat to your birds.

The gadget market offers a host of predator deterrents, including flashing lights, ultrasonic units, calls, urine, decoys, and so on. While these devices may work well in varying scenarios, they should be carefully tested for effectiveness and never used to replace your fundamental methods. We have reduced losses to predators nearly to zero simply through locking our ducks in safe housing by darkness, establishing fenced pasture paddocks, and staggering birds across the homestead according to their vulnerability. Consider that all your efforts to raise birds may be in vain if they are not safe from predator attack.

Nutrition

The importance of a sound feeding program cannot be overemphasized. Nutrition is one of the most neglected phases of management in many home flocks. Nutritional deficiencies and imbalances can be avoided if the guidelines that are outlined here and in chapter 15: Understanding Feeds are implemented.

When planning what to feed your ducklings, always keep these four paramount aspects in mind:

1. You need to provide the right quantity and balance of *all essential nutrients*, including protein, vitamins, and minerals.

2. While an adequate protein level is important, more critical is the quality and balance of amino acids, which are the building blocks of protein.

3. Ducklings have a higher niacin requirement — similar to turkeys, game birds, and broiler-type chicks — than most types of chickens.

4. You should never feed laying or breeding rations with high levels of calcium to growing birds (anything over 1.5 percent calcium should be avoided).

SUGGESTED FEEDING SCHEDULE FOR DUCKLINGS			
Type of Duckling	**0–2 Weeks** Lbs of 18–20% Starter Feed per Bird Daily	**2–7 Weeks** Lbs of 16–18% Grower Feed per Bird Daily	**7–20 Weeks** Lbs of 15–16% Developer Feed per Bird Daily
Broiler	Free choice	Free choice	—
Bantam Breed	Free choice	Free choice 15 min. three times daily	0.15–0.25 (0.07–0.11 kg)
Light Breed	Free choice	Same as above	0.20–0.30 (0.09–0.14 kg)
Medium Breed	Free choice	Same as above	0.25–0.35 (0.11–0.16 kg)
Large Breed	Free choice	Same as above	0.30–0.40 (0.14–0.18 kg)
Muscovy	Free choice	Same as above	0.20–0.40 (0.09–0.18 kg)

Note: The quantity of feed required by ducklings is highly dependent on the availability of natural foods, climatic conditions, and the quality of feed (e.g., birds require larger amounts of high-fiber foods than low-fiber foods to meet their energy requirements).

The rate at which ducklings grow is in direct proportion to the quantity and quality of the feed they consume. For maximum growth they need a diet that provides 20 to 22 percent protein up to two weeks of age and 16 to 18 percent protein from two to twelve weeks.

To stimulate the fastest growth, ducklings should be allowed to eat free choice. If a moderate rate of growth is satisfactory, after two weeks of age, you can limit them to three feedings daily of all the feed they can clean up in 10 to 15 minutes. If feed is not in front of them free choice, it is imperative that sufficient feeder space is provided so that all the ducklings can eat at the same time, without being jostled or bullied by bigger or more aggressive birds. Once healthy ducklings are six weeks old and have access to excellent natural forage, they can be limited to morning and evening feedings daily. Giving the birds their main meal in the evenings will encourage them to forage throughout the day and look forward to coming in at night.

For the first several weeks, small-pelleted (³⁄₃₂ inch) or coarse crumbled feed is preferred; thereafter, larger pellets (³⁄₁₆ inch) will give the best results. Ducklings choke on fine, powdery mash when it's fed dry, and a significant amount of the feed can be wasted. Finely ground feed is better utilized when it is moistened with water or milk to a consistency that will form a crumbly ball when compressed in your hand. A new batch should be mixed up at each feeding to avoid spoilage and food poisoning.

Feeding Programs

The feed program you employ should be designed to fit your situation and goals and most likely will be a combination of two or more of the following options.

Natural

Ducklings, particularly Muscovies, Mallards, and some bantam breeds, that are brooded by their natural mothers are capable of foraging for most of their own ration if there is an abundant supply of insects, wild seeds, and succulent plants. Ponds, lakes, sloughs, marshes, and slow-moving brooks are excellent sources of free food for ducklings. In most situations it is advisable to supply ducklings with concentrated feed for at least the first 10 to 14 days to get them off to a good start.

Grains

Small whole grains such as wheat, milo, kafir, or cracked corn can be fed to ducklings after they are several weeks old. Grains by themselves are not a balanced diet. Ducklings need tender greens and an abundant supply of insects or another protein supplement. They cannot be expected to remain healthy and grow well on a diet consisting exclusively of whole, cracked, or rolled grains. Raw legumes should not be used.

Home-Mixed

It may be practical to mix your ducks' feed if the various ingredients are available at a reasonable price. The formulas given in the tables in appendix A on pages 315–316, Home-Mixed Starting Rations and Home-Mixed Growing Rations, are examples of the types of feed that can be mixed at home. While these rations are not as sophisticated as commercially prepared feeds, they will give good results in most situations if they are mixed properly and the instructions are closely followed. If a vitamin-mineral premix (carried by many feed mills) is used, the rations provided in the table on page 318–319, Complete Rations for Ducklings (Pelleted), can also be home-blended.

Special equipment is not needed to mix duck feed. For small quantities the ingredients can be placed in a large tub and combined with the hands or a shovel. Another method is to pile the measured components in layers on a clean floor and mix with a shovel. A cement mixer or old barrel mounted on a stand and outfitted with a handle, door, and ball bearings can be used for larger quantities. More important than the method used for mixing is that the ingredients are blended thoroughly. During warm weather, no more than a 4-week (3 weeks or less if ground grains are used) supply of feed should be prepared at a time to guard against spoilage and loss of nutrient value.

Commercial

In most localities premixed starter and grower feeds for ducklings are available. When these feeds are used, the instructions should be followed. If rations specifically formulated for ducklings are not available, use the corresponding mixes recommended for game birds, turkeys, or broiler chickens. Regular chick starter/grower rations typically are deficient in niacin but often work satisfactorily if an adequate level of niacin is supplemented.

The question often comes up as to whether medicated feeds should be used for ducklings. There was a time when some medications used in chick feeds were harmful — even deadly — to ducklings. Today the medications commonly used in chick starter and grower do not show negative effects if they are used at the recommended levels. Under many circumstances I still recommend using nonmedicated feeds when possible. However, if nonmedicated feeds are unavailable in your locale, we have tested and found that rations containing amprolium, chlortetracycline, neomycin, oxytetracycline, sulfaquinoxaline, or zinc bacitracin can be used if incorporated in the feed at the proper dosage.

You may want to have a local feed mill mix your feed if enough ducklings are raised to make it practical. The formulas given in the table on page 318–319, Complete Rations for Ducklings (Pelleted), are for complete rations that will provide a balanced diet if mixed properly and not stored for more than 4 weeks (less in hot weather). Ration numbers 9 and 10 are to be fed to ducklings up to two weeks of age and numbers 11 and 12 from two to twelve weeks.

Niacin Requirements

Young waterfowl require two or three times more niacin in their diet than chicks other than broilers (see Niacin Deficiency, page 289). When ducklings are raised in confinement on a ration that is deficient in niacin — such as commercial chick feeds — a niacin supplement should be added to their feed or water. Chick broiler feed typically contains sufficient niacin for growing ducklings.

Niacin can be purchased in tablet form at drugstores and is a common ingredient in poultry vitamin mixes. Adding 5 to 7.5 pounds (2 to 3 kg) of livestock-grade brewer's yeast per 100 pounds (37 kg) of chicken feed (or 2 to 3 cups of brewers yeast per 10 pounds [3.5 kg] of feed) will also prevent a niacin deficiency in ducklings.

Green Feed

The daily feeding of leafy greens to ducklings fortifies their diets with essential vitamins and minerals, provides tasty entertainment, and lowers the possibility of cannibalism. Tender young grass (before it joints), lettuce, chard, endive, watercress, and dandelion leaves are excellent green feeds.

Ducklings will eat their salad only if it is tender and fresh. When greens are placed on the floor of a brooder, they soon wilt and are trampled and soiled. By putting the chopped feed in the ducklings' water trough, you ensure that it remains succulent and clean, and the ducklings spend many contented hours dabbling for the bits of greenery.

Grit

Coarse sand or chick-size granite grit should be kept before ducklings at all times. Grit aids the gizzard in grinding, helping birds to get the most out of their feed. Unless there is a suspected calcium deficiency, oyster shells should not be provided to growing ducklings because this can cause a harmful overdose of calcium.

Feeders

For the first day or two, feed should be placed in containers where the ducklings cannot help but find it. If you have just a few ducklings, you'll find that jar lids, shallow tuna fish cans, and egg flats work satisfactorily. For larger groups make certain the feeders are filled to the top and the feed is obvious. I like to sprinkle a bit of finely chopped tender grass on top of the feed, as this always seems to get their attention.

Feeders (troughs, pans, or bulk hanging feeders) can be purchased or constructed at home. Feeders should be designed so they do not tip over as ducklings jockey for eating space. To keep birds out of trough feeders, a spinner or rod can be attached across the top.

Sufficient feeder space needs to be provided to ensure that each duckling receives its share. If you limit feedings to two or three times daily after the ducklings are several weeks old, be sure to provide adequate trough space so that each bird can eat without having to struggle for its portion of the meal. Once the ducklings are well acquainted with the location of their feed, you can significantly reduce the amount of feed that is wasted by filling troughs or pans no more than one-half to three-quarters full.

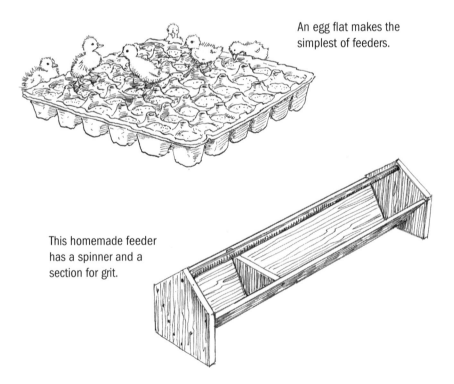

An egg flat makes the simplest of feeders.

This homemade feeder has a spinner and a section for grit.

Creep Feeding

When ducklings are brooded by hens and run with the flock, it is sometimes desirable to feed the young birds separately from the adult ducks. This can be done by devising a creep feeder, where ducklings can be fed their own ration and eat without competing with grown birds.

The basic component of the creep feeder is a panel with openings large enough for only ducklings to pass through and gain access to the feed trough. The doorway panel can be placed across an inside corner or the entrance of a shelter.

Mature ducks can squeeze through smaller holes than is generally realized, and I have found that there is a tendency to make the slots too large. Dimensions for the portals will vary according to the breed raised, and you will probably need to do a little experimenting to find what size works best for your ducks. However, as a general rule, an opening 4 inches (10 cm) tall by 4 inches wide is satisfactory for breeds having mature weights of 6 to 9 pounds (3 to 4 kg). For smaller breeds, such as Mallards, the passageway needs to be approximately 3 inches tall by 3 inches (7.5 cm) wide to keep the old birds from entering and chowing with the youngsters.

Managing Broiler Meat Ducklings

Broiler ducklings are managed to produce the quickest possible growth in the shortest period of time on the least quantity of feed. The breed most often used worldwide for this practice is the Pekin. There are different strains available that grow at different rates and have different fat-to-lean-meat ratios. The strains that have been selected for leaner meat tend to put on weight slightly slower than the fattier and fastest-growing strains. With appropriate management, the fastest and most efficient strains are capable of weighing 8 pounds (3.5 kg) at seven weeks of age on approximately 20 pounds (9 kg) of feed.

Saxony and Silver Appleyards can be used to produce high-quality broiler ducklings that have lower body fat content than most Pekins. Depending on management and ambient temperatures, these are ready to butcher at approximately eight weeks of age.

To stimulate fast growth broiler ducklings must have limited exercise, a continuous supply of high-energy, concentrated feed, and 24 hours of light daily. They are ready to be butchered as soon as their primary wing feathers are developed. If held beyond seven or eight weeks, their feed conversion decreases rapidly (see table on page 209) and they'll commence the molt, making it hard to pick them until they are at least fourteen to sixteen weeks of age.

Producing Lean Ducklings for Meat

A major drawback of raising broiler ducks for meat is the high fat content of their carcasses — especially when raising the fastest-growing strains. While many people enjoy the succulence of quick-grown duckling meat, research strongly indicates that many of us need to reduce the amount of fat in our diets. There are several management practices that can be used to produce leaner ducklings.

One method is to use a high-protein grower ration that contains an energy-to-protein ratio of 65:1 or less. While ducklings fed such rations do not gain weight as rapidly as birds given feed with a wider energy-to-protein ratio (88:1 is normally recommended), the actual amount of meat will be almost identical.

A second method, which is more effective and practical for the small-flock owner, is to raise a breed other than Aylesbury or Pekin; Appleyards, Saxony, Muscovies, and all of the Mediumweight, Lightweight, and bantam breeds are naturally lower in body fat. Then, rather than limiting their exercise and feeding ducklings for the fastest possible growth, they can be allowed to forage in a grassy yard or pasture, which allows them to get considerable exercise while consuming greens and insects. Not only does this lower their fat content, but there is growing evidence that this also changes the composition of body fat to a form that is more healthful for humans. The birds should be given enough feed to keep them healthy and growing well, which usually translates to two or three feedings of concentrated feeds daily in a quantity that they will clean up in 10 to 15 minutes. When butchered at twelve to twenty weeks of age, the fat content of birds raised in this manner is closer to that of wild ducks. In a trial involving 30 Rouen drakes from a production-bred strain, we found that when this method was employed even these medium-large meat birds produced carcasses that were no more fatty than roasting chickens.

TYPICAL GROWTH RATE, FEED CONSUMPTION, AND FEED CONVERSION
of Pekin Ducklings Raised in Small Flocks

These ducklings were fed free choice on a 20% protein, 1,400 kcal/lb ration from 0 to 2 weeks; then fed 16% protein, 1,400 kcal/lb ration from 3 to 12 weeks. They were exposed to 24 hours of light daily and raised in confinement.

Age (Weeks)	Live Weight in lbs (kg)	Feed per Bird in lbs (kg)	Feed per Lb of Bird in lbs (kg)
5	5.07 (2.30)	10.85 (4.92)	2.14 (0.97)
6	6.10 (2.77)	14.52 (6.59)	2.38 (1.08)
7	6.85 (3.11)	18.43 (8.36)	2.69 (1.22)
8	7.43 (3.37)	22.36 (10.14)	3.01 (1.37)
9	7.87 (3.57)	26.52 (12.03)	3.37 (1.53)
10	8.36 (3.79)	31.43 (14.26)	3.76 (1.71)
11	8.58 (3.89)	36.47 (16.54)	4.25 (1.93)
12	8.43 (3.82)	40.55 (18.39)	4.81 (2.18)

Sexing Ducklings

Over the years I have been told many secret methods for determining the sex of day-old ducklings. When tested, these techniques have proven to be only 50 percent accurate at best. To my knowledge the following methods are the most trustworthy procedures for sexing live ducklings.

Vent Sexing

The only sure way to sex ducklings of all breeds before they are five to eight weeks old is by examining the cloaca. While it is easier to vent-sex waterfowl than land fowl, this procedure still requires practice, an understanding of the bird's physiology, and care to avoid permanent injury to the bird. Too often, persons attempt to sex ducklings without first acquiring the needed skills, injuring the birds or wrongly identifying the gender.

At best, a written account on sexing ducklings is a poor substitute for a live demonstration by an experienced sexer. I highly recommend that you have a knowledgeable waterfowl sexer show you how to vent-sex ducks before trying it yourself.

While ducks of all ages can be sexed by this method, I suggest that novices practice on birds (one of the medium to large breeds) that are approximately a week old. At this age ducklings are easier to hold and their sex organs are large enough to be readily identified. As ducks get older the sphincter muscles that surround the vent become stronger, and often it is more difficult to expose the cloaca. Extremely young ducklings are usually harder for the beginner to hold for sexing, and the sex organs are so small that they can be tough to identify until a person knows what to look for. There are three points to keep in mind when vent sexing ducklings.

1. Young ducklings are extremely tender, and clumsy fingers can kill the bird or permanently injure it.
2. A bird has not been sexed until the cloaca has been inverted. People often assume that if they can't find a penis after applying a little pressure to the sides of the vent, the bird is a female. However, a duckling cannot be identified as a female until the cloaca has been inverted and no penis is evident.
3. The sex organs of ducklings are tiny, so it is essential that birds are sexed in excellent light.

You should be able to sex ducklings if the following steps are carefully implemented. There are several methods for holding birds for sexing, but I'll describe only the way I find most comfortable.

Step 1. Hold the bird upside down with its head pointed toward you. If the duck is large, its neck and head can be held between your legs.

Steps 1 and 2

Step 2. Use the middle and/or index finger of your right hand to bend the tail to a position where it is almost touching the bird's back.

Step 3. Push the fuzz or feathers surrounding the vent out of the way so you can see what you are doing.

Step 4. Place your thumbs on either side of the vent and apply pressure down and slightly out.

Step 5. Use the index finger of your right hand to apply pressure down and out on the backside of the vent. This inverts the cloaca and exposes the sex organs.

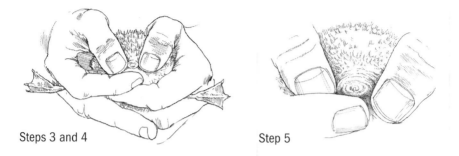

Steps 3 and 4   Step 5

If It's a Drake

If the duckling you are sexing is a male, a corkscrew-shaped penis will pop up near the center of the cloaca (you must look carefully, since sometimes only the tip of the penis will be visible). In drakes only a few days old, the penis is extremely small and can appear almost translucent. By the time males are a couple of weeks of age, their sex organs are easier to see because of their larger size and deeper white or yellowish color.

If It's a Female

Should you have a female, no penis will be visible when the cloaca is inverted. In ducks that are more than several weeks old, the female genital eminence is often visible as a small, dark (usually gray) protuberance that resembles an undeveloped penis.

If You Have Difficulty

The most common problem encountered by inexperienced sexers is getting the cloaca inverted. If this is your problem, make sure that:

- You have the bird's tail bent back double as in the illustration
- You are applying pressure down and out simultaneously with both your thumbs and your right-hand index finger

As with most skills, your speed and accuracy in sexing will improve with practice. If you get discouraged on your initial attempts, wait a few days until you have regained your confidence and then try again.

VENT-SEXING THE OLDER BIRD

1. A convenient method for restraining large ducks for sexing is to turn the bird upside down and hold its neck between your legs.
2. After bending the tail back (see below), push aside the feathers surrounding the vent, position your thumbs on the sides and your index finger on the back of the vent, and invert the cloaca by applying pressure down and out.
3. The cloaca and penis of an eight-week-old drake are shown below. If the cloaca is inverted sufficiently to expose two dimples and still no penis is evident, it can be assumed that the bird is a female.

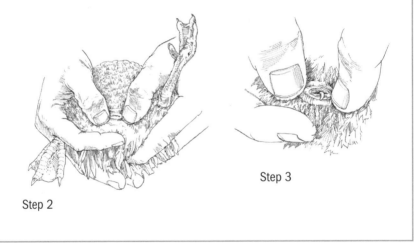

Step 2

Step 3

Sexing by Voice

By five to eight weeks of age, ducks of most breeds can be distinguished by their voices. To sex birds by this method, catch each duckling individually, and as it protests its predicament, listen carefully for the female's distinctive harsh quack. The young male's vocalization is intermediate between its baby and adult voices, resembling an elongated *wongh*. Due to the structure of the voice box, drakes are incapable of producing the hard quack sounds of the female.

Sexing by Bill Color

It is possible to sex purebred ducklings of some varieties by their bill color at an early age. Day-olds of Snowy and Harlequin colored varieties can often be sexed by bill color with at least 75 to 95 percent accuracy. Drakelets typically have a fairly uniformly colored grayish or greenish bill, while the bills of ducklets are yellowish with a dark tip. In wild-colored varieties such as Gray Calls, Mallards, and Rouens, the bill of a drake normally turns dull green by the age of five to eight weeks, while the female's turns dark brown with varying amounts of orange shading. Females in most other colored varieties have a darker bill than the drake by five to eight weeks of age. Determining gender by bill color is unreliable with some crossbred ducklings as variation in bill color can simply be a sign of genetic differences.

Sexing by Plumage

Drakes do not acquire their adult nuptial plumage until they are four to five months old. However, in many colored varieties, there are sufficient variations in the juvenile feathering of drakes and females that make it possible to differentiate the sexes of ducklings that are five to eight weeks old. In drakes of wild-colored varieties such as Gray Calls, Mallards, and Rouens, the feathers of the crown of the head and back often are darker with varying amounts of green iridescence and have less brown penciling than ducks.

Marking Ducklings

Several methods can be employed to mark individual birds, including the use of leg or wing bands and the notching and perforating of the webbing between the duckling's toes.

Bands and toe punches are available from poultry-equipment suppliers and feed stores. If leg bands are applied to growing ducklings, great care must be taken to use rings that are large enough to allow good circulation, and they should be changed frequently as the legs grow.

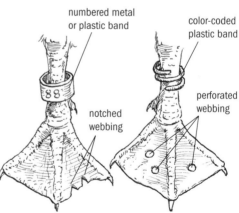

Marking birds for easy identification

A REMARKABLE RESCUE

One morning my wife, Millie, found a twelve-day-old duckling wedged in the lip of a water fountain. The only sign that the soaked bird was still alive was a barely perceptible movement of a leg. Millie put the bird in a small bucket of lukewarm water and carefully held him there with his head above the surface. In 5 minutes the little fellow was able to hold his head up. Millie toweled him off and then, using a handheld hair dryer set on low, dried his down, being careful not to overheat him. We placed the duckling in a 90°F (32°C) battery brooder with a group of five-day-old ducklings. An hour later the "miracle" duckling was up and running around with no sign of his near-death experience.

Brooding Ducklings with Other Birds

Many people who have duck flocks also raise other poultry. Consequently, sometimes ducklings and other young fowl are being brooded at the same time. To save space and equipment, it is tempting to brood the various species together.

We routinely brood ducklings and goslings together, which has the added advantage of making it more likely that as adults the two species will get along amicably. Especially if the brooder facility is not overcrowded, people frequently will brood young waterfowl and land fowl together. Do keep in mind, however, that baby ducks, geese, quail, guineas, chickens, and turkeys have unique habits, temperaments, growth rates, and management requirements that can increase risks when they are brooded in mixed groups. Land fowl chicks have a tendency to pick on ducklings and goslings. Ducklings and goslings naturally have wetter habits, which can create problems for land fowl. As a rule, ideally, waterfowl and land fowl are brooded separately.

14

Managing Adult Ducks

ADULT DUCKS DO NOT REQUIRE a lot of time-consuming care or specialized facilities. By using common sense and being observant, novices can manage ducks successfully.

Basic Guidelines

The most common error made with ducks is trying to raise them as if they were chickens. For good results, the unique habits and needs of each species of fowl must be recognized and provided for. The following are absolute musts for long-term success in duck keeping.

1. Think like a duck! Humans are the top predator in the world, and our natural inclination is to be straight-line thinkers. Every truly successful duck raiser I have observed over the years has had the ability to step out of a predator mentality and has learned how to look at the world from a duck's perspective. One way to do this is to take time to sit and observe the daily goings-on of ducks and what makes your ducks feel safe in your setting.

2. Provide an environment in which the ducks feel safe. Your job when working around your flock is to move and work in a way that does not startle them or make them feel that they are about to be devoured. Remember: Ducks recognize us as huge predators, with eyes on the front of our faces and long fingers that they perceive as claws. Averting your eyes, walking in a relaxed manner, and talking reassuringly can lessen your birds' anxieties.

3. Protect your birds from predation. Ducks raised in small flocks seldom die from disease or exposure to severe weather, but significant numbers are lost to both domestic and wild land, air, and aquatic predators. In most locations ducks should be penned *every* night in an enclosure that predators cannot dig, climb, or fly into. The enclosure should also be free of openings that dexterous predators such as raccoons can reach through to grab terrified birds.

4. Supply a balanced diet in an adequate quantity, and make any dietary changes gradually to avoid the possibility of lethal enteritis. (Enteritis is the inflammation of the digestive tract which can cause diarrhea and in severe cases may result in death.) Following close behind predators, improper nutrition is the second leading cause of problems encountered by duck keepers.

5. Provide a steady supply of drinking water. Ducks do not thrive if they are frequently left without water during daylight hours. (We have experimented with not having water at night for mature ducks. While egg production is slightly reduced, it is a great aid in keeping the night quarters drier.)

6. Furnish suitable living conditions. Keeping ducks locked up in yards covered with deep mud and stagnant waterholes is an invitation to trouble.

7. Do not disturb them more than necessary. Waterfowl thrive on tranquillity.

8. When catching ducks, walk them into a corner as quietly as possible, using a long-handled net or your hands to pick them up. Because the legs and feet of ducks are easily injured, do not grab or carry them by their legs.

Housing

Because ducks have the finest down jackets available, they require less protection from inclement weather than do chickens. Ducks often prefer to stay outside in most weather. If predators are not an issue and shade is available, in mild climates it is possible to raise ducks without artificial shelters. A good windbreak — straw bales work well — will provide sufficient protection to keep ducks comfortable in regions where temperatures occasionally dip below freezing. A more substantial shelter is needed in areas where extremely low temperatures are common.

Because waterfowl normally perch on water or the ground at night, the main reason for housing ducks in mild weather is to protect them from nocturnal predators (see appendix C, Predators, page 326). A tight fence at least

4 feet (1.25 m) tall, with no gaps at ground level, enclosing the night yard will keep most dogs and coyotes out. Both dogs and coyotes have a strong tendency to try to crawl under or through fences, rather than exposing themselves and jumping over the top — although there are exceptions. We have used 5-foot- (1.5 m) tall deer-and-rabbit fencing (with vertical stays every 6 inches [15 cm] and graduated horizontal wires starting at an inch [2.5 cm] apart at the bottom and getting farther apart toward the top) with good success. No-climb horse fence (with 2" × 4" [5 × 10 cm] openings) can also be used; however, it is more difficult to install, and ducks will occasionally get their heads or necks entwined and stuck in it.

In areas where thieves such as raccoons, foxes, skunks, opossums, bobcats, domestic cats, bears, or wild boars are present, for good protection it is necessary to mount strands of electrified wire on insulators on the outside of the above fence. The first wire should be 2 to 3 inches (5 to 7.5 cm) from ground level, the second one approximately 6 inches (15 cm) higher, and one strand at the very top. An additional strand should be approximately 2½ feet (0.75 m) from the ground, where bobcats, bears, and wild boar frequent.

MASKED BANDITS

Their clownish countenance notwithstanding, raccoons are the most destructive of all fowl thieves in many regions of North America. Adaptable to both rural and urban settings, smart, persistent, strong, amazingly dexterous, and able swimmers, this large member of the weasel family is aptly labeled "superpredator."

Raccoons will climb up and over the tallest fence as if it were a ladder put there for their convenience. And with their nimble paws, they have been known to reach through wire netting with 1-inch- (2.5 cm) wide openings, grab dozing or panicked birds, and kill them through the fence.

The only sure way to keep ducks safe from these nocturnal hunters is to lock the waterfowl every night in a building or covered pen (one that has solid walls for at least the bottom 30 inches [75 cm]) or a tightly fenced yard that is protected by properly installed electrical netting or fencing. (See page 224 for fencing ideas and appendix C, page 326, for additional information on predators.)

Here in the foothills of western Oregon, we have an abundance of predators, including raccoon, fox, skunk, mink, opossum, coyote, bobcat, and cougar. For several years we have been testing 4-foot- (1.25 m) high electric poultry netting to protect large waterfowl day and night — with a 100 percent success rate against terrestrial predators.

Where cougars and large owls frequent, ducks should be enclosed in a covered pen or building. Weasels and mink are notorious for being able to squeeze through seemingly impossibly small openings; therefore, to protect poultry from them, an ultratight shelter is required.

If you do not have an empty building that can be converted into a duck house, an inexpensive shedlike structure can be constructed. In the illustration on page 220, a practical duck house is shown. This type of shelter can be made portable and may be used for either baby or adult birds.

THE TRIPLEX DUCK RUN

Here in Western Oregon we have long, wet winters; dry, windy summers; and numerous predators — most notably raccoons, fox, skunks, mink, opossums, coyotes, bobcats, cougars, eagles, and large owls. If ducks are left unprotected, sooner or later they will become some critter's dinner. Furthermore, with all the ducks we have, they would demolish grassy runs if allowed outside during the wettest times of the year. Over the years we have found this triplex duck run ensures safety for the birds, provides them a healthy environment, and protects the grass.

The triplex duck run comprises three main components:

1. An area inside a building that can be tightly closed at night against all predators and that provides a minimum of 2 to 6 square feet (0.25 to 0.5 sq m) per bird, depending on their size and temperament. This area is bedded with cedar or pine shavings, coarse sawdust, or straw.

2. A bedded outside yard providing an additional 2 to 10 square feet (0.25 to 1 sq m) per bird, to which ducks have access during the day. To prevent mud we cover these yards with 3 to 4 inches (7.5 to 10 cm) of coarse, round gravel, topped with 2 to 3 inches (5 to 7.5 cm) of sand, covered with a layer of coarse cedar sawdust. To reduce the

Since ducks are short, there is no need to have the walls of the duck house be more than 3 feet (1 m) high at the lowest point, except for your convenience in gathering eggs and cleaning out old litter. By having the nests attached along the outside and outfitted with a hinged top, eggs can be gathered without entering the shed. Three solid walls and a wire front are recommended except in regions with severe winters, where it is better to have a closed front. Good ventilation is essential, however, even in cold climates, because ducks will fare poorly if they are forced to stay in stuffy, damp quarters.

Dirt, sand, wood, or cement floors can be satisfactory in duck buildings. To keep predators from burrowing into buildings with dirt or sand floors, it may be necessary to place a wire, wooden, or cement barrier around the outside perimeter of the structure.

amount of maintenance needed to keep the yards in good condition, over the years we have covered most of them with roofs.

3. A grassy yard, which is irrigated during the summer, is available to the ducks during nice weather. Shade is supplied by shrubs, trees, or overhead grape arbors. We try to supply a minimum of 50 square feet (4.5 sq m) per bird.

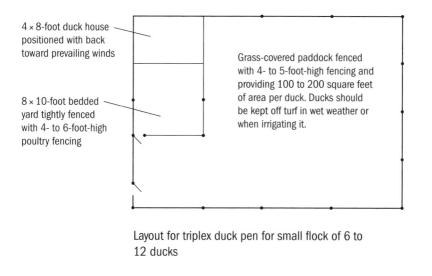

4 × 8-foot duck house positioned with back toward prevailing winds

8 × 10-foot bedded yard tightly fenced with 4- to 6-foot-high poultry fencing

Grass-covered paddock fenced with 4- to 5-foot-high fencing and providing 100 to 200 square feet of area per duck. Ducks should be kept off turf in wet weather or when irrigating it.

Layout for triplex duck pen for small flock of 6 to 12 ducks

To ensure good drainage, place the shed on a slope or build the floor up 4 to 8 inches (10 to 20 cm) above ground level with dirt, sand, gravel, and/or bedding. If the duck house is located in a low spot, water will accumulate during wet weather, making it impossible to keep the litter in good condition.

When medium- to large-size ducks are housed only at night, allow at least 2.5 to 6 square feet (0.25 to 0.55 sq m) of floor space per bird. If you anticipate keeping your ducks inside continuously during severe weather, a minimum of 6 to 8 square feet (0.5 to 0.75 sq m) of floor space per bird should be provided. Bantam breeds require approximately half the floor space of larger ducks.

The attached nests on this practical duck house make egg gathering easier.

Cold-Weather Housing

Ducks are well prepared to remain comfortable in freezing weather with their thick garb of feathers and down. Nonetheless, feed consumption can be reduced and egg production increased if ducks are protected from the severity of bitter northern winters, particularly at night, when the birds are inactive.

Insulated Quarters

In locations where temperatures are below freezing for extended periods of time, insulating the duck house can reduce feed consumption and increase egg production. If the duck house has double-wall construction, air spaces can be filled with sawdust, chopped straw, or a commercial insulation. A traditional way to insulate shelters is to stack straw bales against the outside walls from ground level to the roof.

Deep Litter

With the deep litter system, when the bedding becomes sufficiently soiled to warrant refreshing, a thin layer of fresh bedding is added on top. This accumulation of bedding in the duck house keeps ducks off cold floors and reduces the penetration of cold from the ground during periods of low temperatures. The deep litter system actually goes one step further by producing heat through decomposition of the litter and manure. This system has been tested in cold climates and has been found to increase winter egg production by up to 20 percent. It also eliminates the need to clean out duck houses more than once or twice yearly and produces excellent organic matter for the home garden.

To employ the deep-litter method, follow these three steps: (1) cover the duck house floor with 2 to 4 inches (10 to 15 cm) of dry bedding; (2) periodically stir the bedding, then add a fresh top layer of material to keep the surface in good condition; and (3) clean out the old litter each spring and/or fall. Because the litter may accumulate to a depth of 18 to 30 inches (45 to 75 cm) or more, the walls of duck houses where this system is used need to be tall enough to take this accumulation into account.

The Duck Yard

Many duck raisers have a fenced yard where birds are locked in at nighttime or continuously if space is limited. The ideal yard has a minimum of 10 to 25 square feet (1 to 2.25 sq m) of ground space per duck, natural or artificial shade, a slope that provides good drainage, and no stagnant water holes or deep mud.

If the soil on the land where you live has poor drainage, the duck yard should be covered with pea gravel, sand, straw, or wood shavings, with the center of the yard built up higher than the outside edges. When choosing a location for your duck yard, keep in mind that a dense population of birds can kill young trees because of the high nitrogen content of manure. The area around the perimeter of yards should be kept free of briar patches, low-growing bushes and trees, and tall grass or other hiding places for predators. Around our duck yards we keep a swath at least 10 feet (3 m) wide mowed or grazed low to discourage surprise attacks by fowl-loving meat eaters. It is fine to have trees or shrubs in duck yards for shade and the edible fruit or seeds they produce, but remember that woody plants that grow on the outside and overhang fences provide ready access for climbing predators.

RAISING DUCKS WITH CHICKENS

When not overcrowded, ducks can be raised with chickens and other species of birds. However, because ducks have wet habits and wash their bills and faces in drinking water, it is best to have separate drinking water for the chickens on an elevated perch out of the reach of ducks. Also, there are diseases that different species of birds can carry and spread to one another. Commercially, ducks and chickens (and other land fowl) should be kept separate.

Keeping Ducks on Wire Floors

Adult ducks can be kept in houses or hutches with wire-covered floors or in elevated, all-wire cages. The main advantage of this system is that birds can be maintained in extremely limited space without problems of sloppy litter or muddy yards. Some disadvantages are that feed consumption can be higher since the birds cannot forage, foot problems can develop (particularly among the larger breeds), hatchability of eggs is often reduced, predators can harass or injure birds from underneath, and life is less enjoyable for the ducks.

To prevent lameness wire mesh used on floors should not have openings larger than 1 inch × ½ inch (2.5 × 1.25 cm). However, if finer mesh is used, manure may not pass through, causing buildup problems. Plastic poultry flooring that is gentler on the feet can also be used. A minimum of 2 square feet (0.25 sq m) of floor space per medium- or large-breed duck should be allowed, with double or triple this amount being recommended when feasible.

Bedding

Ducks can withstand wet weather, but they should never be forced to stay in muddy, filthy yards or buildings. The floors of buildings should be covered with several inches of mold- and toxin-free litter such as sand, sawdust, wood shavings, peanut hulls, peat moss, crushed corncobs, flax, or straw. In muddy or snow-covered yards, mounds of bedding 4 to 6 inches (10 to 15 cm) thick should be provided to give ample space where all ducks can roost comfortably. Add new layers of clean bedding as needed.

Shade

Ducks must have access to shade in hot weather. They suffer if forced to remain in the sun at temperatures above 70°F (21°C). When ducks are confined to a yard lacking natural shade, a simple shelter should be provided that will supply adequate shade for all birds residing there. Feed troughs can also be located under this cover to protect the feed from exposure to the elements.

A simple shelter provides shade and protects feed from sun and rain.

Nests

Clean eggs hatch better and retain their freshness for eating longer than eggs that are heavily soiled. Having adequate nests for your ducks will help produce clean eggs and reduce the chances of eggs being broken. To allow ducks to become familiar with them, install nests at least 2 weeks before you expect the first eggs. One nest for every four or five females usually suffices if eggs are gathered daily.

For medium to large ducks, nests approximately 1 foot square (0.1 m) and 12 inches (30 cm) high are recommended. Duck nests are placed on the ground and do not need solid bottoms if sufficient nesting material is provided to keep nests off hard or dirty floors. If nests are inside buildings, they do not require a top. Keeping nests well furnished with clean, dry nesting material such as wood shavings or straw will encourage ducks to use them and result in fewer broken or soiled eggs.

CHOOSING FENCING MATERIAL

The main purpose of fences is to keep ducks in and predators out. Most breeds of ducks can be kept in with a 2- to 3-foot- (0.5 to 1 m) high barrier if the openings are small enough and the bottom of the fence is tight and close to the ground, preventing them from squeezing underneath.

However, to keep dogs and coyotes out, tight fences at least 4 to 5 feet (1.25 to 1.52 m) tall are necessary. (In general, dogs and coyotes will not jump over a 5-foot fence into unfamiliar territory, although there is no guarantee they never will.) Our farm is surrounded by coyotes, but we have never had one breach our 5-foot-high perimeter fences. Remember: Dogs, coyotes, and ducks prefer to go under or through a fence rather than over its top.

The following is a brief review of some types of fencing material:

1. **Wire chicken netting.** This lightweight material commonly comes with either 1- or 2-inch (2.5 or 5 cm) openings and is relatively inexpensive. Raccoons have been known to tear through it; posts need to be used every 4 to 6 feet (1.25 to 1.75 m) to keep the bottom tight; and ducks will sometimes get their bills snagged in it. A strand of smooth wire stretched tight and attached to the chicken wire along the bottom edge will greatly enhance the effectiveness of this fencing.

2. **Deer and rabbit woven fencing.** A moderately heavy-duty fencing with vertical stays that are 6 inches (15 cm) apart and horizontal wires that start out 1 inch (2.5 cm) apart, gradually increasing to 4 inches (10 cm) apart at the top. It is generally available in heights from 4 to 8 feet (1.25 to 2.5 m). Properly installed, this wire will stay tight for decades. It is the primary type of wire we use on our farm for perimeter fences.

3. **Woven field fencing.** A moderately heavy-duty fencing usually used for sheep, hogs, and cattle. It has vertical stays that are either 6 or 12 inches (15 or 30 cm) apart and horizontal wires that start out at 2½ to 3 inches (6.5 to 7.5 cm) apart, gradually increasing to 5 inches (12.5 cm) or more at the top. It will last a long time, but most ducks can walk or squeeze through the openings.

4. **Welded wire.** A strong, rather rigid wire that comes with openings of various dimensions, including ½ × 1 inch (1.25 × 2.5 cm), 1 × 1 (2.5 × 2.5 cm), 1 × 2 (2.5 × 5 cm), and 2 × 4 (5 × 10 cm). Heights range from 2 to 8 feet (0.5 to 2.5 m). We use ½ × 1 inch and 1 × 1 meshes for floors of water porches and water platforms (areas where the birds drink), and 2-foot-high 1 × 2-inch mesh for divider fences in breeding and brooding yards (these short fences are easy for humans to step over). In general we do not use wire with 2 × 4-inch openings because ducks will occasionally get their heads and wings caught in them.

5. **Chain-link fencing.** A heavy-duty, very strong, expensive wire product that lasts a lifetime under many applications. It comes in many heights and is excellent for strong perimeter fences if installed properly.

6. **Electric fencing.** When installed properly, a few strands of electrified wire work almost like magic to keep out climbing predators. Typical installation uses a minimum of three strands of aluminum or steel wire mounted on 2- to 6-inch- (5 to 15 cm) long insulators on the outside of the fence. The bottom strain runs 2 to 4 inches (5 to 10 cm) above the ground, a second strand runs 18 to 24 inches (45 to 60 cm) above the ground, and the third runs along the top of the fence. **Caution:** For safety, always use a livestock fence–charger to energize electric fences!

7. **Electroplastic fencing.** As long as large predators do not "bulldoze" into this lightweight electric netting, it does a great job of keeping birds in and predators out. It comes in heights ranging from 22 to 48 inches (55 to 120 cm) and can be readily moved. Electroplastic fencing can be a lifesaver when installed around a setting duck that has stationed her nest outside a protective yard or building. (See appendix G for sources.)

8. **Game-bird netting.** This netting comes with openings of different sizes and in rolls of many widths and is excellent for covering pens and yards where winged predators such as owls, hawks, and eagles are problematic. (See appendix G for sources.)

Portable Panels

Lightweight panels constructed from 1 × 2-inch or 1 × 4-inch lumber and wire netting frequently come in handy. They can be used to divide pens, for isolating ducks that are setting or have young, as a catch pen in a large yard, and for small pens that can be moved around to utilize grass in different areas. To retain most breeds of ducks, the panels only need to be 2 or 3 feet (0.5 to 1 m) high. Panels that will be used to pen young ducklings should be covered with netting that has openings no larger than 1 inch × 1 inch (2.5 × 2.5 cm).

Farm and ranch supply stores often sell galvanized stock panels that are 16 feet (5 m) long. We use dozens of these handy panels around our farm. The 34-inch- (85 cm) tall "hog" version is handy for making temporary pens. The 52-inch- (132 cm) high style (choose panels with small openings at the bottom so ducks cannot walk though them) will keep out most dogs and can be used for gates or around the perimeter of duck yards.

Swimming Water

Ducks enjoy swimming, and they add charm and interest to lakes, ponds, and brooks. Where natural bodies of water do not exist, earthen or cement ponds can be constructed. Be sure to design them for easy drainage and cleaning. To keep earthen banks from eroding, we use riprap (large rocks). A 4- to 6-inch- (10 to 15 cm) deep layer of round gravel around the perimeter of small ponds is helpful in keeping ducks from tracking in mud and drilling with their bills.

Under many circumstances, rather than digging a pond, it may be advantageous to use children's wading pools, masonry mortar pans (available at hardware stores), or large dishpans. They are easy to empty, they can be moved to fresh ground every day to prevent mud, or they can be placed on top of a wire platform or a bed of gravel.

Contrary to popular belief, ducks can be raised successfully without water for swimming. In fact, there are some advantages to having only drinking water. If a small pond is overcrowded, sanitation problems often arise. Ducks must not be allowed to swim in or drink filthy, stagnant water.

Drinking Water

Ducks need to have a regular supply of reasonably clean drinking water. For top egg production, water should be provided day and night. If ducks have been conditioned to not have drinking water at night at least 6 weeks prior to onset of lay, egg production is minimally affected by withholding drinking water for up to 9 hours each night. For this procedure to work, consistency is paramount! However, if you have separate nighttime quarters for your ducks, the pen will stay drier if water is available only in the daytime portion of the pen.

For a few ducks, drinking water can be supplied in a dishpan, a vacuum-style waterer (available at most feed stores), or other appropriately sized container. Never use a container that is proportionally tall and slender where a duck can get stuck upside down and drown. Accidental drowning can occur when the body of a duck fits snugly into the container. For larger flocks there are many choices of automatic waterers, ranging from nipple systems (if a nipple system is the only source of water, ducks typically are dirty and unkempt) to ones with different sizes of weight-activated bowls. We have also used bowl-type automatic horse waterers that provide water a couple of inches deep with good success.

If no bathing water is available, ducks will keep their nostrils and eyes cleaner if drinking water is deep enough to allow them to submerge their heads. Because ducks frequently wash their bills, it is impossible to keep drinking water clear; however, water should be kept reasonably clean and should never be allowed to become putrid.

To keep unhealthful mud holes from developing around watering areas, waterers can be placed on a platform covered with wire mesh or plastic poultry flooring. For most breeds openings of ½ × ½ inch, ½ × 1 inch, or 1 inch × 1 inch (1.25 to 2.5 cm sq) work well. Manure tends to pass through the larger holes more readily, but if you raise bantam ducks and you do not want them to stick their bills through the flooring, it is better to use one of the two smaller sizes. Availability and the particulars of your situation will dictate the suitable size of the opening.

For additional protection a pit 12 to 24 inches (30 to 60 cm) deep can be made underneath the platform and filled with gravel. A simple method to prevent ducks from drilling in the soil around the waterer is to place a 3- to 5-foot- (1 to 1.5 m) square piece of welded wire mesh or plastic poultry flooring on the ground underneath the waterer (if using this method, the waterers and wire will need to be moved to new locations as needed to prevent buildup).

During cold weather when there is soft snow, ducks can sustain themselves for a time by eating the icy crystals. However, it is preferable to give them drinking water at least twice daily when the temperature is cold enough to freeze all available liquid.

OUR BREEDING SCHEDULE

Because we are working with many breeds and strains of ducks, we use a large number of relatively small breeding pens. Matings consist of pairs, trios, pens (one drake to three to five ducks), and flocks containing a ratio of one male to four to six ducks. Our breeding season is January to June; therefore, our schedule is designed to maximize production during these months.

1. Prior to making up the breeding pens, the ducks are blood tested for pullorum-typhoid.

2. At least 4 weeks before hatching eggs are saved, ducks are placed in their breeding pens and artificial lights are used to lengthen the days to approximately 13 hours of light daily (we use the minimum artificial light possible to stimulate the egg production and fertility desired).

3. We provide one nest box for every one to four females. To encourage exercise, feeders are located at the far end of the pens inside the duck barns, while the waterers are located 15 to 30 feet away in the outside runs.

4. Three weeks before hatching eggs are desired, ducks are switched (over the course of several days) from a 14 percent–protein maintenance ration to our 19 percent–protein breeder pellets.

5. For large breeds that may have reduced fertility, we provide bathing pans or in-ground ponds, which stimulate mating in more lethargic birds.

6. During wet weather the ducks have access to inside shelters and covered runs. When the grassy yards are sufficiently dry so that the ducks will not muck them up, they are allowed outside during the day.

7. They are locked up every night in the duck barns to provide predator protection.

8. As soon as the laying season is over, the birds are switched from the breeder ration to a maintenance ration with 0.6 to 1 percent calcium and 14 percent protein.

Raising Ducks on Salt Water

Ducks can be raised successfully in marine areas. Most ocean bays and inlets are teeming with an abundant supply of plant and animal life that ducks relish. Because domestic ducks have a lower tolerance for salt than do wild sea ducks, fresh (nonsaline) drinking water should be supplied at all times. Birds that are raised for meat on marine waterways may need to be confined in a pen or yard and fed a grain-based diet for at least 3 to 4 weeks prior to butchering to avoid fishy-flavored meat.

Nutrition

Feed is the single most expensive item in raising ducks, normally representing 60 to 80 percent of the total cost. Finding ways to save on feed expenditures will significantly decrease the cost of your duck project. However, no matter how inexpensive, poor-quality feeds that do not meet all the nutrient requirements of your birds will not lead to savings — and in the long run will derail the success of your duck venture.

For top egg production, ducks must be fed an adequate amount of concentrated feed having 16 to 20 percent crude protein. Three weeks before the first eggs are expected and throughout the laying season, a laying ration should be fed free choice or the amount the birds will clean up in 10 to 15 minutes should be fed two or three times daily (see the table on page 231 for recommended daily feed allowances). Ducks lay the best when they are in a semifat condition, but excessive fatness is harmful and must be avoided. A sudden change in the diet of ducks that are laying often results in a sharp decline in egg production, and it can throw them into a premature molt from which it may take at least 10 to 16 weeks to recover.

While not in production, mature ducks can be given a maintenance ration containing 13 to 14 percent crude protein and fed just enough to keep them in good condition. The quantity of feed required by birds is highly affected by various factors, including weather and the caloric density of the feed. During cold periods ducks must eat considerably more feed than when temperatures are moderate.

Fish by-products such as fish meal are excellent sources of high-quality proteins and are used in some commercially prepared poultry feeds. These ingredients are excellent for immature birds and adult breeders that produce hatching eggs, but laying ducks that supply eating eggs must be given feeds that are low in (less than 4 percent), or do not contain, fish products to prevent the production of strong-flavored eggs.

Feeding Programs

In formulating a feed program for your adult duck flock, try to find ways to obtain a high degree of productivity while using food resources that normally go unharvested.

Natural

The least expensive way to feed ducks is to make them forage for their food. This option is limited to situations where wild and natural foods are in abundant supply and when top egg production is not important. It must be remembered that the quantity of natural feed fluctuates widely during the various seasons of the year. While there may be periods when ducks can find most and possibly all of their own feed, there will probably be times when most of their food will need to be supplied in the form of grain and mixed rations. Ducks must never be allowed to deteriorate and become thin due to lack of feed. Negligence in this area can permanently damage their productivity.

---

10 COMMON CAUSES OF POOR EGG PRODUCTION

1. Sudden change in appearance, texture, flavor, or content of feed
2. Birds unexpectedly left without feed or water
3. Sudden or periodic changes in hours of daily light exposure
4. Decreasing length of daylight
5. Birds periodically being startled (a sense of well-being and safety is paramount)
6. Improper diet or inadequate quantity of feed
7. Heavy infestation of internal or external parasites
8. Overcrowding or excessive number of drakes in flock
9. Filthy living quarters
10. Contaminated water or moldy feeds

SUGGESTED FEEDING SCHEDULE
FOR ADULT DUCKS

Size of Duck	Holding Period When Birds Are Not Producing Lbs of 12–14% Protein Feed*	3 Weeks Prior to and During Laying Season Lbs of 16% Protein Feed*	2 Weeks Prior to Butchering Mature Ducks Lbs of High-Energy Feed*
2–3 lbs (1–1.5 kg)	0.15–0.25 (0.07–0.11 kg)	0.20–0.30 (0.09–0.14 kg)	Free choice
4–5 lbs (1.75–2.25 kg)	0.20–0.30 (0.09–0.14 kg)	0.30–0.40 (0.14–0.18 kg)	Free choice
6–7 lbs (2.75–3.25 kg)	0.25–0.35 (0.11–0.16 kg)	0.40–0.50 (0.18–0.23 kg)	Free choice
8–9 lbs (3.5–4 kg)	0.30–0.45 (0.14–0.2 kg)	0.45–0.60 (0.2–0.27 kg)	Free choice
Muscovy	0.30–0.40 (0.14–0.18 kg)	0.40–0.60 (0.18–0.27 kg)	Free choice

*Per bird daily

Note: The quantities of feed that you need to supply to ducks are highly dependent on the availability of natural foods, the climate conditions, and the quality and caloric density of feed (e.g., birds require larger amounts of high-fiber foods than low-fiber foods to meet their energy requirements).

Grains and Legumes

Whole grains, by themselves, are not a complete poultry feed. However, if ducks are given a protein concentrate or can forage in waterways or pastures, a supplement of grain will often satisfy their dietary needs, particularly when the birds are not in production.

Ducks are not as fond of oats and barley as they are of grains such as wheat, milo, and corn. But oats and barley are good waterfowl feeds, and the birds will learn to eat them if made to do so.

Corn is a high-energy grain and an excellent feed for cold weather and for fattening poultry — if you desire fat meat. During hot weather the diet of ducks should not consist of more than 25 to 50 percent corn. If waterfowl are fed too much corn during exceptionally hot weather, egg production drops and health problems can arise.

Raw legumes should not be fed to poultry. However, heat-processed soybeans are used extensively as a protein and energy source in duck feeds.

Home-Mixed

It may be practical to mix your own feed if the various ingredients are available at a reasonable price. Mixing a home ration for ducks is simpler than for chickens. All ingredients for chicken feeds need to be finely ground; otherwise the birds will pick out the grains and leave the finer particles. This problem is usually not encountered with ducks since they tend to scoop up their feed rather than picking it up one kernel at a time.

The formulas in the table in appendix A, Home-Mixed Rations for Adult Ducks (see page 317), are examples of the types of feeds that can be mixed at home for small flocks of ducks. If a commercially prepared vitamin:mineral premix is used, the rations provided in the table on pages 320–321, Complete Rations for Adult Ducks (Pelleted), can be prepared at home as well. Similar products that are more readily available in your area can be substituted for the suggested ingredients in the tables.

Commercial

The simplest way to provide ducks with a balanced diet and stimulate top production is to purchase commercially mixed concentrated feed from a reliable manufacturer. A good ration is formulated to ensure a proper balance of carbohydrates, fats, proteins, vitamins, minerals, and fiber.

In many areas feeds manufactured specifically for breeding and laying ducks are not available. Often chicken, turkey, and game-bird feeds can be used in place of duck mixtures. (Chicken rations usually need to be supplemented with additional niacin. See Niacin Requirements on page 205.) Ducks waste less feed if they are fed pellets rather than crumbles or mashes.

If you are raising a large number of ducks, it may be economical to have a local feed mill custom-mix your duck feed. In one of the tables in appendix A, Complete Rations for Adult Ducks (Pelleted), I give feed formulas that we have used successfully for a number of years (see pages 320–321).

Growing Duck Feed

It is possible to raise a good portion of the feed required by a home flock of ducks in a surprisingly small area. In parts of the country where field corn thrives, this grain yields large quantities of ears that can be harvested and shelled by hand or broken in two and thrown to the ducks to shell for themselves. Grains such as wheat and rye provide fall and spring grazing, and when the seeds are mature, the ducks can be allowed into the patch for a short time daily to harvest their own feed. Grain sorghums such as kafir and milo produce large seed heads that can be hand-harvested.

Our favorite perennial plants for producing significant quantities of food and shade in our duck yards include mulberries, apples, and grapes (the grapes are grown on a rectangular four-posted arbor that is 8 feet [2.5 m] tall). The fruit and seeds from these plants require no harvesting because they fall to the ground when mature and the ducks eagerly consume them.

A large assortment of natural crops is available for planting in and around bodies of water. Some of the favorites of ducks include wild celery, wild rice, wild millet, small bulrush, smartweed, and chufa tubers. Addresses of nurseries specializing in natural game-bird crops can be found in hunting and outdoor magazines. Ducks also enjoy and benefit from vegetable and fruit produce that cannot be used in the kitchen.

Pasture

While ducks can be raised in barren yards or pens, they enjoy and benefit significantly from succulent and appropriate vegetation when it is available. Good-quality forage can lower their feed consumption and lessens the possibility of some dietary deficiencies. Ducks cannot eat mature pasture (except the seeds), so it should be mowed occasionally to encourage new growth.

Orchards are an excellent location for duck pastures. The vegetation can help crowd out noxious weeds and provides a protective and attractive ground cover. Ducks can significantly reduce the numbers of harmful insects in orchards and help clean up diseased and windfall fruits.

There are a variety of grasses and legumes that make good permanent pasture for ducks. The main requirements are that the plant thrives in your region and that it produces succulent forage that your duck finds palatable. We have used forage-type perennial ryegrass, fine fescue, Kentucky blue, bent, and orchard grass and New Zealand white, Ladino, and White Dutch clovers in our pastures in various combinations with good success. For quick and succulent forage, we have used annual ryegrass, which has the added benefit of producing a large enough seed head that the ducks readily eat it. (Modern lawn-type grasses are sometimes inoculated with fungal endophytes to increase their drought tolerance and hardiness, but the effect of this organism on ducks is unclear.) Your local agriculture Extension service specialist can give advice on what varieties do well in your climate.

POTATOES ARE GREAT DUCK FOOD

Carol Deppe, author of Breed Your Own Vegetable Varieties: The Gardener's and Farmer's Guide to Plant Breeding and Seed Saving *and* The Resilient Gardener: Gardening in Good Times and Bad, *shares some of her experiences feeding ducks.*

I feed substantial amounts of potatoes to my flock of 32 free-range Ancona ducks. Potatoes can replace all the corn in the duck ration as well as provide a good bit of the protein. We grow potatoes to eat and to sell. Growing potatoes means we have prime potatoes and seconds. We use the latter as duck food. They make such great duck food and save so much on the duck chow bill that we are expanding our potato planting in part to provide duck food during the fall and winter months.

I feed only cooked potatoes to ducks. Raw potatoes aren't food for monogastric animals, as they contain inhibitors for many digestive enzymes, including those that digest starch. Cattle, sheep, and goats can be fed raw potatoes, as their ruminant bacteria deactivate the inhibitors. But for poultry, cook the spuds.

A second proviso — potatoes can be a substantial part of the food for a *free-range* flock. I don't advise using large amounts of potatoes with the *confined* flock without reformulating the feed to accommodate the potatoes.

The potatoes I use for duck food are misshapen, small, or have healed splits. Any green potatoes get culled in the field at harvest time. Green potatoes have elevated levels of glycoalkaloids and are unsafe for human or duck consumption. Potatoes with any green should be discarded entirely. (Elevated levels of glycoalkaloid are in the entire potato, not just the skin or green part.)

To prepare potatoes for my ducks, I put about three gallons of culls plus a few for my own dinner in a big kettle of water. Occasionally a small potato floats, and I discard it. These sometimes represent potatoes the plant started to make, but then robbed nutritionally for the potatoes it finished. (They can have higher glycoalkaloid content.) It takes about 30 minutes to bring the salted water in my big pot to a boil and another 45 minutes to cook the potatoes. When potatoes come to a boil and cook that slowly, they generally don't break up or slough. I usually get my dinner out of the pot too, so the labor involved isn't any more than for most dinners. With the flock of 32, I prepare potatoes every day or every other day in season, and that is enough to give the birds about 1½ to 2 gallons [5.5 to 7.5 L] of spuds per day. This amount is approximately as much as the flock wants in a day, given both potatoes and broiler chow free choice. I used to mash the ducks' potatoes a little, but it isn't necessary. I do cut the top layer in the bowl the first time or two in the season that I feed potatoes.

I feed the ducks smorgasbord style, with chicken broiler chow (20–22 percent protein) free choice in one bowl and high-carb food such as corn, potatoes, or cooked squash in another. Two additional buckets contain oyster shell and granite grit. The birds mix and match according to their nutritional needs, the forage quality, and the weather. On mild maritime winter days when there are lots of high-protein goodies such as nightcrawlers, slugs, and sow bugs, my flock eats little of the commercial chicken broiler chow, but good amounts of the high-carb food. On days below freezing, there is no fauna about for the ducks, so they eat substantial amounts of both broiler chow and high-carb food. When the weather is cold, the birds eat more carbs than when it is warm. Laying ladies eat mostly oyster shell grit and just an occasional mouthful of granite grit. Drakes, ladies not laying, and ducklings eat the granite grit, not the oyster shell. Ducks seem to be good at evaluating their nutritional needs and choosing food accordingly.

When my ducks have potatoes in the high-carb bowl instead of corn or squash, they eat only about half as much broiler chow as usual. Feeding potatoes to my ducks saves the entire corn bill and about half of the broiler chow bill for the entire season in which I have potatoes available. Overall, feeding potatoes free choice along with broiler chow reduces my feed bill by roughly two-thirds during the potato season. My ducks really like the spuds. If I give them a bowl of broiler chow, a bowl of spuds, and a bowl of commercial corn, they simply don't eat the corn. They would much rather have the potatoes. On those mild winter days when there are lots of worms and other delicious high-protein fauna about, my ducks eat mostly just forage, potatoes, and grit. They eat hardly any of their high-protein broiler chow at all. The delight of ducks for potatoes was great news for me. I can't grow broiler chow. I *can* grow potatoes.

Feeding Insects to Ducks

Insects are a bountiful source of high-quality protein. While ducks of all ages are accomplished bug hunters, their consumption of winged insects can be increased by burning a low-wattage bulb 18 inches (30 to 45 cm) above ground level in the duck yard at night. As the insects swarm around the light, the birds have hours of leisurely dining.

Using Surplus Eggs and Milk

Milk and hardboiled eggs were a traditional source of high-quality protein for poultry in a bygone era. I have had many old-timers tell me that when they supplemented their poultry with milk and/or eggs, their birds grew faster, had glossier plumage, and laid more eggs. However, if you decide to try either of these, keep in mind that there are potential problems — such as spreading disease from contaminated eggs.

The only practical way to feed liquid milk to ducks is to mix small quantities into a dry ration. When milk is given in pans or buckets, ducks play in it, covering themselves and the surrounding area with the sticky stuff. Under these conditions, sticky feathers, eye infections, and food poisoning can develop.

Eggs should always be hard-boiled before they are given to poultry, and if fed to birds that are going to lay, they should be chopped or mashed to an unrecognizable form. Feeding raw eggs can result in a biotin deficiency and may increase the chances of birds eating their own eggs. To avoid spoilage, milk and eggs should not be mixed with the duck ration until feeding time.

Feeding Leftovers

Kitchen and garden refuse must be fed to producing birds in moderation. Leftovers normally are high in starch, fiber, and liquid and low in most other nutrients. Ducks cannot lay well if their minimum requirements for protein, minerals, and vitamins are not being ingested. Only provide an amount that the birds will clean up before it spoils.

Grit and Calcium

To digest their feed to the best advantage, ducks need to have a supply of granite grit, coarse sand, or small gravel. Approximately two weeks prior to and throughout the laying season, laying ducks need to have their diets supplemented with calcium if they are going to lay strong-shelled eggs. Most manufactured laying feeds contain the correct amount of calcium, but when grains or home mixes are used, a calcium-rich product such as dried, crushed egg shells; oyster shells; or ground limestone should be fed free choice.

Feeders

When adult ducks are fed a limited quantity of feed, sufficient feeder space must be allowed so that all the birds can eat at one time without crowding. Otherwise, less aggressive and smaller ducks may be pushed aside and not get their full share.

Approximately 4 to 6 linear inches (10 to 15 cm) of feeder space should be allowed per bird. Ducks can eat from both sides of most feeders, so a trough 5 feet (1.5 m) long provides 10 linear feet (3 m) of feeder space, enough for a flock of 20 to 30 birds. V-shaped troughs, with spinners mounted along the top to keep birds out, are easily constructed and provide sanitary eating conditions. Hanging, freestanding or wall-mounted bulk self-feeders are available in many sizes. Depending on the size of your flock and the size of your bulk feeder, it is possible to fill these and have sufficient feed available for an extended period of time. In some situations a significant disadvantage of these bulk feeders is that wild birds and rodents can consume large quantities of feed. (I have had people tell me that their ducks were eating over 2 pounds of feed per day per duck — which is physically impossible! Once necessary precautions were taken, feed consumption returned to the expected 0.25 to 0.4 pounds [0.1 to 0.2 kg] per bird per day.)

DURABLE CONTAINERS

Flexible rubber pans (with sides approximately 4 inches [10 cm] high) are commonly available in various sizes at feed and farm-supply stores. These receptacles do not crack when they freeze and make excellent feeders and water/bathing pans that last for years under normal use.

Finishing Roasting Ducks

When ducks have been forced to forage for a significant portion of their feed during the growing period, they can be fed grain or finishing pellets free choice for 2 or 4 weeks prior to butchering. If they have ranged widely, it can be advantageous to pen them in a restricted yard (allowing a minimum of

10 to 25 square feet [1 to 2.5 sq m] per bird) during the finishing period. This time of heavy feeding and restricted exercise will encourage the ducks to gain weight rapidly, resulting in a larger carcass and succulent, tender meat.

The Laying Flock

When managing birds that are laying, it is important to keep in mind that producing eggs is extremely hard work, and it taxes a female's body to produce these marvelous but complex orbs — which have all the essentials for life. No matter the species, for a bird to lay well and produce the most nutritious eating eggs or viable hatching eggs, there are four key essentials: the bird must feel safe and content, have a steady supply of reasonably clean drinking water, consume a diet that contains all the essential nutrients in adequate quantities, and live in an environment that is healthy and is conducive to egg production.

Although it depends on breed and the environment they are kept in, most female ducks of nonbantam breeds lay their eggs before 9 AM (bantam ducks have more of a tendency to lay throughout the morning). To ensure that eggs are not lost in fields or bodies of water, it is a good practice to confine laying ducks to a yard or building at night. Drakes are not required for non-fertile (eating) egg production, but their presence can enlarge the foraging range of a flock.

Lighting Needs

Light is one of the primary controllers of reproduction in birds. Increasing or decreasing day length, total hours of daily light, and light intensity are all factors. Increasing day length signals the ducks' reproductive system to start "gearing up" for sperm and egg production. In nature, as soon as the days start lengthening after the winter solstice, hormone levels in both the male and the female start changing to support reproduction. Conversely, in the fall and winter months, decreasing day length and falling temperatures signal the birds' reproductive systems to "shut down." Ducks are even more sensitive to light than are chickens — meaning that they respond more quickly to slight increases or decreases in day length.

When top egg production is desired during the short days of fall, winter, and early spring months, ducks, like chickens, must be exposed to artificial lighting. This is especially true the farther you are from the equator and the colder your winter. Due to ducks' greater sensitivity to light, ducks require less light than chickens do to stimulate and maintain egg production under most circumstances.

My rule of thumb is to utilize as little artificial light as possible to get the desired level of egg and sperm production. By using the minimal amount of light possible, if egg production unexpectedly starts to decline, you have room to "bump up" the light stimulus by increasing day length in 5- to 15-minute increments. In general I have found little or no benefit in providing more than 16 hours of light daily. In fact, once you get past 15 hours of daylight, many ducks have a tendency to get broody and want to set on eggs rather than lay.

The milder the temperature, the shorter the day length needed to stimulate egg and sperm production. In general drakes need a longer day length to produce good fertility than females need to start egg production. In our situation, we have found that 12 hours of daylight is sufficient — along with temperatures that stay above 20°F (–7°C) and an egg-supporting diet — to stimulate egg production. For good fertility we find 13 hours of light daily to be advantageous in drakes six to eighteen months of age. As drakes get older they tend to need a longer day length (typically 14 to 15 hours) for good fertility.

It is essential that the length of daylight never decreases while ducks are producing, or the rate of lay may be severely diminished. Even a reduction of only 15 to 30 minutes per 24-hour period for several days can negatively affect heavily producing hens. If you use an automatic time switch that turns lights on before daybreak and off after nightfall, ducks can be provided the desired hours of light daily.

TYPICAL EFFECTS OF MANAGEMENT ON EGG PRODUCTION

| | Annual Egg Production per Duck | | |
Treatment	Domestic Mallards	Commercial Rouens	Khaki Campbells
Fed whole or cracked grains; given access to pasture and pond; exposed to natural day length	25–40	50–65	75–150
Fed 16% protein laying pellets; given access to pasture; exposed to natural day length	50–75	75–100	175–225
Fed 16% protein laying pellets; given access to pasture; exposed to 14 hours light daily	85–125	125–150	275–325

To prevent premature egg production, the amount of light young ducks are subjected to needs to be watched carefully. When young ducks are exposed to excessive light or increasing day lengths between the ages of eight and sixteen weeks, they may begin to lay before their bodies are adequately mature. Premature laying (before eighteen to twenty weeks of age) can result in a shortened production life, smaller and fewer eggs, and greater possibility of complications such as prolapsed oviducts.

The intensity of light required to stimulate egg production is relatively low (approximately 1 foot-candle at ground level, the equivalent of 1 incandescent bulb watt per 4 square feet [0.4 sq m] of floor area when the bulb is at a height of 7 or 8 feet [2 or 2.5 m]). When ducks are confined to a building or shed at night, one 40- to 60-watt incandescent bulb 6 to 8 feet (1.75 to 2.5 m) above ground level will provide adequate illumination for each 150 to 250 square feet (14 to 24 sq m) of floor space. In outside yards, one 100-watt bulb with a reflector per 400 square feet (37 sq m) of ground space is recommended.

We use the following lighting schedule with good success for ducks that are hatched in March through July, and I recommend it if you desire maximum efficiency and production. This schedule is specifically designed to induce spring-hatched ducks to start laying during the decreasing day lengths of fall. Especially in very mild climates closer to the equator or if trying to stimulate egg production in the spring, shorter day lengths (approximately 12 hours) are typically sufficient.

LIGHT REQUIREMENTS

Age	Hours of Light Daily
0–7 weeks	24 hours to keep flock calm at night
8–18 weeks	Natural day length
19 weeks	Add 15 minutes of light weekly until 13 to 14 hours of total light (natural and artificial) is reached. If the day is naturally 13 to 14 hours or longer, add no artificial light. Once natural light falls below 13 to 14 hours, add enough light morning and evening to provide 13 to 14 hours. If egg production declines significantly, add 15 minutes of light weekly until egg production is satisfactory or a maximum of 15 to 16 hours of light daily is reached. Maintain a constant level until ducks stop laying or they are force-molted.

Identifying Producing Ducks

Shortly before production commences, and throughout the laying season, the abdomens of ducks swell noticeably. Ducks that are in production can be identified by their large, moist vents and widely spread pubic bones. (Ducks tend to be sensitive to being handled as they approach and are in production, so catching and handling during this time is not generally recommended.) As the season progresses the bright yellow bills of higher-producing, white-plumaged ducks will fade to pale yellow. In ducks with colored plumage, the bill normally darkens during the laying season.

Encouraging Ducks to Use Nests

To encourage ducks to lay where you want them to, place dummy eggs in nest boxes several weeks before the beginning of the laying season. While nest eggs can be purchased, I have found that turning them out on a lathe is an enjoyable rainy-day project. You do not have to worry about making them exactly the correct shape and size. Wooden eggs that are painted white or beige have the advantage of being more visible to ducks. They also last longer and are easier to clean.

Production Life of Ducks

With good management Lightweight breed pullets begin laying at seventeen to twenty-four weeks, while pullets of the larger breeds typically commence laying sometime from twenty to thirty weeks. Ducks lay the greatest number of eggs in their first year of production but normally show only a minor decrease in productivity during the second and third year, when eggs are larger in size. Many ducks will lay some eggs until they are five to eight years old.

Breaking Up Broody Ducks

When ducks become broody, their egg production falls off or ceases. To induce broodies back into production, isolate them in a well-lit pen without nests or dark corners and provide drinking water and feed. In these surroundings ducks normally lose their maternal desires in 3 to 10 days and can be returned to the flock. Because broodiness seems to be contagious, remove ducks from the flock promptly when they show the first signs of wanting to set.

IDENTIFYING PROBLEMS IN THE LAYING FLOCK

Symptom	Common Cause
Thin- or soft-shelled eggs	Usually a vitamin D_3 or calcium deficiency; also high temperatures or abnormal reproductive organs
Eggs decrease in size	A gradual decrease in size is common as the laying season progresses. However, hot weather, excessive environmental stresses, and dietary deficiencies accentuate the problem.
Odd-shaped eggs	Temporary malfunction of or abnormal or diseased oviducts
Pale-yellow yolks	A diet lacking carotene, which is supplied by products such as yellow corn and green plants. Ducks fed wheat-based rations without access to pasture or greens usually produce anemic-colored egg yolks.
Bright orange-red yolks	Diets high in corn and/or green feeds
Blood or meat spots in egg interior	Internal hemorrhage; eggs fit for eating, but some people find these distasteful
Blood on shell exterior	Ruptured blood vessel or injured cloaca. Frequently occurs when young ducks begin laying.
Back of head and neck bare of feathers; in extreme cases, skin is lacerated	Too many drakes, which results in excessive mating activity. Drakes sometimes have favorites.
Ducks go into a premature molt	Sudden diet or lighting change; fright; birds frequently left without water; onset of hot weather
Eggs lost in bodies of water, pasture, or hidden nests	Ducks not locked in a pen or small yard at night, or are turned out before they lay

Force-Molting

Normally, ducks lose their feathers yearly and replace them with new ones. Ducks ordinarily do not molt while they are laying since their bodies cannot support the work of growing feathers and forming eggs at the same time. Under natural conditions this arrangement works out fine; the duck lays in the spring, hatches and broods her young until they are able to fend for themselves, and then molts, with her new garb being ready for the fall and winter months.

When ducks from prolific strains are managed for high egg yields, they typically commence laying at sixteen to twenty weeks of age and lay without long interruptions for 12 months or longer. After this extended lay cycle, their feathers are worn and egg quality normally declines. Because it is practical to keep laying ducks for 3 years or more under some circumstances, it can be beneficial to force-molt them every 10 to 18 months to give their bodies a rest from laying and to allow them to grow new feathers. After this time of rejuvenation, ducks typically lay larger eggs during their second lay cycle.

The most common time to force-molt ducks is when egg production is naturally decreasing and temperatures are not going to be bitterly cold. The time it takes ducks to molt and begin laying again varies, but it is generally from 8 to 12 weeks.

Laying birds are thrown into a molt by sudden changes in their diet and environment. The following schedule is suggested as a simple method for force-molting. (Make necessary adjustments to your specific situation.)

FORCE-MOLTING SCHEDULE

1st day:	Discontinue artificial lighting and remove all water and feed.
2nd and 3rd days:	Provide drinking water but no feed.
4th day:	Commence feeding again, but substitute the laying ration with whole oats fed free choice.
15th day:	In addition to the oats, supply in a separate feeder 0.2 to 0.3 pounds (0.09 to 0.14 kg) per bird (depending on the size of bird) of a 16 to 18 percent–protein duck, waterfowl, or poultry grower, or game-bird flight conditioner ration once daily.
42nd day:	Gradually start replacing oats and grower feed with layer ration, and supply 13 to 14 hours of light daily. (If natural light is longer than 13 to 14 hours, add no additional light.)

Selecting Breeders

To maintain the productivity of ducks from one generation to the next, choose breeders carefully. Select specimens possessing robust health, strong legs, and no inheritable deformities. If purebred ducks are raised, breed characteristics for typical size, shape, carriage, color, and markings should also be considered.

Characteristics to Avoid

The following deformities and weaknesses can be inherited and will occasionally show up in ducks. Keep in mind that many of these defects can also be the result of injury, nutrition, inadequate incubation, or improper environment. Unless the deformity is due to an extraneous cause, birds with any of these faults should not be kept for reproduction.

Blindness. Certain strains of poultry, particularly some that are highly inbred, show a significant incidence of clouded or white pupils. Noninheritable blindness can be caused by injuries such as being poked, pecked, or burned by hot surfaces (such as heatlamps) or caustic substances (such as ammonia and hydrated lime).

Crossed bill. When a duck has this defect, the upper and lower mandibles are not aligned. In my experience, most crooked bills are the result of ducklings cutting their bill on the eggshell at the time of hatching, which then grows out crooked; this is not inheritable.

Deformed feet or curled toes. The feet or toes are bent so severely that the duck has difficulty walking (most ducks have slightly curved toes). Deformed feet and toes can be caused by genetic defect, malnutrition, or from an inadequate incubation environment (such as improper turning, improper egg positioning, and improper or fluctuating temperature).

Neck deformities. Necks that are distinctly crooked can be the result of a genetic defect, inadequate incubation environment, or injury. Ducks, especially Runners, can develop severely kinked necks if they have insufficient headroom in their cage or transportation carrier. Crested ducks are more likely to exhibit atypical necks than are other breeds.

Roach back. A deformed spinal column causes a humped, crooked, or shortened back. Deformed spinal columns are most often the result of a congenital defect.

Scoop bill. Scoop bills have an unnaturally deep, concave depression along the topline of the bill. Ducks with this condition often have poor vigor.

Weak legs. This problem can often be traced to nutritional deficiencies, obesity, or overexertion, but there are inheritable forms.

Wry tail. Rather than pointing straight back as is normal, wry tails are constantly cocked to one side. While a wry tail can be caused by malposition in the egg, if I know the incubation environment has been good, I assume any hatchings that hatch with this defect carry a genetic fault. However, if a duck has a normal tail and suddenly develops wry tail, I have found it is usually caused by trauma to the spinal column.

Neurological abnormalities. Neurological symptoms can be triggered in ducklings by inadequate oxygen levels in incubators or during transport, or by ear infection or head trauma in birds of all ages. There are also inheritable forms of neurological defects that cause birds balance problems or atypical head position. Crested ducks are particularly susceptible to neurological defects (see page 71 for more information on Crested ducks).

Catching and Holding Ducks

Handle ducks carefully, since their wings and legs are easily injured. Their legs are especially susceptible to injury; therefore, avoid running them on rough ground or where they will trip over feed troughs and other obstacles. I highly recommend walking waterfowl as calmly as possible into a V-shaped corner or a small catching pen rather than having a wild chase around a large yard. If ducks are struggling to stay on their legs and are helping to propel themselves by beating their wings against the ground, you are pushing them too hard! You need to back off immediately and let the birds calm down. A long-handled fish or game-bird net can help minimize trauma when catching ducks.

HOW TO CATCH AND HOLD

A safe and convenient method for catching ducks is to grasp the bird gently by the neck.

Small and medium-size ducks can be caught and held with a thumb over each wing and the hands encircling the body.

Here are two methods for holding ducks that are comfortable and safe for both the bird and the holder.

Remember: *Never* grab ducks by their legs. Depending on the size of the duck and the circumstances, we use several techniques to pick ducks up. Small and medium-size ducks can be picked up and held with a thumb over each wing and the hands encircling the body. Another method we use with Light-weight and bantam ducks is to hook or encircle our thumb and forefinger around the base of the wings (where they join the body). When picked up using this method, most ducks quit struggling and relax; your free hand can then slide palm-up underneath the bird to support its weight, while controlling the legs by hooking your thumb and finger around them. Larger birds can be grasped securely but gently by the neck or with one hand over each wing to subdue them; then slide a hand under the breast and secure the legs by hooking your thumb and index finger around them. Talking reassuringly and gently stroking its back will help calm the bird.

My favorite way of carrying a duck is to have its weight resting on my forearm with its neck or body secured between my body and arm. When the

TINY TINA

Tiny Tina was my favorite boyhood duck. She would come running when I called her name and loved to follow me around the home place when the duck-yard gate was left open. We would hike across the adjoining 80-acre field, and when it was time to return, she would fly home. Occasionally, I would take her to the creek so she could have a dip and hunt water bugs.

Tiny was a dandy broody, and each spring for 6 years she laid a nest full of eggs and hatched them. The 7th year she laid a few abnormally small eggs and showed no interest in incubating them. Her voice started sounding hoarse, and I was concerned that she was sick.

That summer when she molted, to my astonishment her head became mottled with greenish black and her sides turned gray. And there in the middle of her tail were several partially curled drake feathers! That fall Tiny Tina's name was officially changed to Tiny Tim. This was my introduction at any early age to the fickle phenomenon of hormones! I came to learn that occasionally, due to various causes, normal hormone levels can change radically — causing a bird to undergo a partial sex change.

wings and feet are held gently but securely, the duck likely will not be injured or escape. This method also minimizes the chances of you being soiled.

Special precautions should be taken when handling Muscovies. They are surprisingly powerful and typically have long, sharp claws. When handling any type of bird, hold it away from your face to avoid any injury to your eyes. Wearing gloves and long-sleeved garments helps avoid cuts and scratches.

Distinguishing the Sex of Mature Ducks

The gender of adult ducks can be identified by secondary sex characteristics such as voice, feather formation, plumage color, and body size. The voices of drakes resemble a hoarse "wongh," in contrast to the quacking of females. Except during the molt, drakes typically display several curled tail feathers, and nonwhite varieties are usually more colorful than females.

Muscovies are practically mute, although the females can quack weakly. Drakes lack the curled tail feathers of true ducks, and the two sexes are similar in color. However, the sexes can be readily identified by the drakes' larger size and the females' smaller facial skin patch.

Clipping Wings

To keep flying ducks from wandering, it may be necessary to clip their wings. This is easily and painlessly accomplished by cutting off the 10 primary feathers of one wing with scissors. So that birds do not look unbalanced, the two outermost primaries can be left intact. Once a year ducks molt their wing feathers and replace them, so to keep them grounded, trim their wings annually.

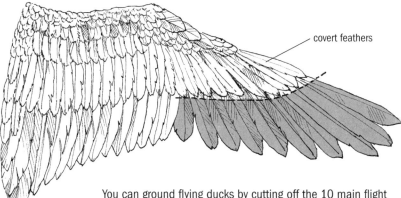

covert feathers

You can ground flying ducks by cutting off the 10 main flight feathers (the portion that is highlighted) of one wing. Cut directly below the tips of the covert feathers.

Reducing Manure Shoveling

If you plan ahead, you can reduce the amount of manure that needs to be shoveled by allowing the birds to spread much of their droppings directly onto the land where it is needed. Fallow parts of the garden can be used as the duck yard, by fencing off harvested portions or using movable duck tractors. Covering the ground with 2 to 6 inches (5 to 15 cm) of organic matter such as cornstalks, leaves, or straw will not only keep the ducks out of mud but also will protect the dirt from being packed down and provide an excellent environment for earthworms and soil-enriching microorganisms.

At the appropriate time the partially decomposed bedding and manure can be raked up and composted or worked directly into the soil. In gardens where this method has been used, lush crops of many kinds of vegetables and fruits have flourished with minimal use of commercial fertilizers.

15

Understanding Feeds

TO BE HEALTHY AND PRODUCTIVE, a duck must consume a diet that provides all the essential nutrients in the proper quantity and in the correct balance with other nutrients. Because ducks are one of the hardiest and most adaptable of all domestic fowl, people often make the mistake of assuming that they can thrive on a poor diet. What a duck eats has a profound influence on its growth rate, appearance, productivity, and longevity. If fed an improper diet, even ducks out of world-class stock will not perform well or look like their parents. Cheap feeds can be expensive in the long run because of their low caloric density or low-quality essential nutrients. Appendix A (page 315) provides reference for mixing your own rations.

Main Dietary Stages of Ducks

Wild ducks are continually adapting their diet to the special requirements of a particular season. Under domestic conditions, it is the responsibility of duck keepers to make sure their birds are consuming an appropriate diet for each of the six main stages of life.

1. **Starting stage.** For the first two weeks, ducklings require the highest level of protein and some vitamins. Starter feed must be in a form that ducklings can consume and easily digest. Recommended protein levels are 18 to 20 percent with a calcium-to-phosphorus ratio of approximately 1:1 (the calcium level should not be greater than 1 percent).

2. **Growing stage.** In weeks three to eight, ducklings are one of the fastest growing of domestic fowl. To sustain such rapid growth and minimize physical deformities, ducklings must consume a high-quality diet that

meets all their nutritional requirements but at the same time does not push them beyond a reasonable growth rate. For pets, breeding stock, and exhibition birds, a ration with 15 to 16 percent high-quality protein will encourage a moderate growth rate. For maximum growth rate, a protein level of 18 percent is often used. The calcium-to-phosphorus ratio should be approximately 1:1 (again, the calcium level should not be greater than 1 percent). Be aware that the higher the protein percentage, the higher the likelihood of physical problems, such as wing and leg deformities and kidney and liver damage.

3. **Developing stage.** Depending on the breed and management, ducks have reached 70 to 90 percent of their adult weight by the time they are nine weeks old. Between nine and twenty weeks of age, they grow slowly, replace their juvenile plumage, and mature sexually. During this developing stage, a level of 13 to 14 percent high-quality protein is adequate with a calcium-to-phosphorus ratio of approximately 1:1 and a maximum calcium level of 1 percent.

4. **Laying stage.** Three weeks prior to and throughout the laying season, the requirements for protein, many vitamins, and some minerals (notably calcium) increase markedly. If the protein is high quality and the essential amino acids are carefully balanced, 15 percent crude protein is adequate. However, 16 to 17 percent protein is often used, and 18 to 20 percent protein can result in larger eggs (as temperatures increase, protein levels need to be raised if egg size is to be maintained). Calcium requirements jump to 2.5 to 3 percent, while total phosphorus levels stay at about 0.65 percent.

5. **Holding or maintenance stage.** When fully mature ducks are not laying, their protein requirements drop to 12 to 14 percent with a calcium level of 0.6 to 1 percent and total phosphorus at 0.6 percent.

6. **Molting stage.** Each year mature ducks typically go through an eclipse and a nuptial molt. Growing feathers require additional protein (15 to 16 percent high-quality protein is usually adequate) and other nutrients; therefore, a well-balanced diet during a molt will encourage healthy plumage. The addition of animal protein (5 to 10 percent by volume cat kibbles is a good source) and 10 to 20 percent oats (by volume) during the molt can produce wonderful feather quality.

Types of Feeds and Supplements

When you go to a feed store, the many different kinds of feeds available can be confusing. To further complicate matters, different feed companies use different names for the same class of feed. Also, in some parts of North America, feed stores may not carry "duck feed." Here's a guide to help you understand common feeds and how to use them.

Duck or Waterfowl Starter/Grower

These feeds are formulated specifically for the dietary needs of baby ducks and geese. Because starter and grower feeds are similar, most feed companies make a combination starter/grower rather than two separate rations. These feeds typically have 18 to 20 percent protein.

If you are raising your ducks for meat and want the maximum growth rate, feed these as the sole ration. However, if you are raising your ducks for pets, breeding stock, or exhibition, and your duck starter/grower has more than 16 percent protein, the birds will normally live longer and have fewer leg and wing problems if their growth rate is slowed down by adding oats. Begin by adding 5 percent oats by volume (either meal, rolled, whole, or pelleted) the first week and an additional 5 percent each week until the birds are receiving three parts starter/grower and one part oats.

FEED FOR SPECIAL NEEDS

Always make sure you provide a diet that is age- and life-stage appropriate. For example, ducklings need a ration formulated specifically for baby waterfowl, whereas ducks that are laying require a diet that meets their special needs.

Chick Starter/Grower

Most feed stores sell chick starter/grower. These feeds are formulated for egg-type baby chickens, which have lower niacin requirements than ducks. If you use chick starter with ducklings, supplemental niacin should be supplied. Feed manufacturers can vary the amount of niacin they include in their feeds, but under most circumstances, satisfactory results are achieved if you add 100 to 150 mg of niacin per gallon of drinking water from zero to eight weeks of age for ducklings being fed chick starter/grower. Niacin is available in tablet or powder form at health food and drugstores. Keep in mind that excessively high levels of niacin can be toxic, so do not be tempted to put in extra for "good measure."

Meat Bird or Broiler Starter/Grower

Broiler starters are formulated for Cornish-cross meat chickens, which have niacin requirements similar to ducklings. Some feed manufacturers also sell feeds under various names that are formulated to meet the requirements of all species of meat-type birds, including ducks, geese, turkeys, and broiler chickens. If they are nonmedicated or medicated with a drug that is not harmful to ducklings (such as amprolium or zinc bacitracin), these feeds can be used for ducklings. However, they are excessively high in protein (usually 20 percent or more) for pet, breeding, and exhibition ducks, so they should be supplemented with oats, as outlined under Duck or Waterfowl Starter/Grower.

Turkey or Game-Bird Starter/Grower

Turkey and game-bird starter/growers have plenty of niacin for ducklings but are excessively high in protein at 22 percent or higher. Because of their high protein content, they can cause a variety of physical problems (especially leg and wing deformities) and damage the kidneys and liver if fed as the sole ration to ducklings. Some people have satisfactory results by mixing one part chick starter, one part turkey or game-bird starter/grower, and one-half part uncooked oatmeal for a starting and growing ration for ducklings when duck starter/grower is not available.

Duck Developer

Developer rations usually contain about 14 percent protein and are designed to keep ducks between nine weeks of age and sexual maturity healthy and in good feather condition without becoming obese. If a duck developer is not available from your feed store, you can use game-bird flight conditioner or mix your own by blending together five parts duck starter/grower (broiler starter/meat builder), two parts oats, two and a half parts wheat, and one-half part cat kibbles.

Game-Bird Flight Conditioner

This feed is formulated to help game birds be in good feather and muscle condition for flying. With ducks it is useful as a developer, maintenance, and exhibition feed. A duck-developer and duck-maintenance ration can be made by mixing four parts game-bird flight conditioner with one part oats.

COMMON CAUSES OF INCREASED FEED CONSUMPTION

1. Fast growth rate
2. Onset of egg production
3. Decreasing environmental temperatures
4. Wind chill
5. High-fiber feed with low caloric density
6. Heavy infestation of internal parasites
7. Increased exercise
8. Decline in natural foods (grass, seeds, insects, invertebrates)
9. Rodents or wild birds stealing feed

Layer Rations

Layer rations are formulated for birds that are producing eating eggs. (If eggs are going to be hatched, a breeder ration should be used.)

To lay well and not deplete the body of minerals (especially calcium) and vitamins, ducks need to eat a laying feed 2 to 3 weeks prior to and throughout the laying season. Most good chicken laying rations work satisfactorily for ducks. While conducting feed trials with laying ducks, we had a Khaki Campbell lay 357 eggs her pullet year on a 16 percent–protein chicken layer manufactured by a national feed company.

Warning: Because of the high calcium levels (2.5–4.5 percent Ca) of laying feeds, they should never be used for growing birds. More than 1.2 percent calcium in the diet of nonlaying birds can cause permanent damage to organs and the skeleton, or even death. When possible (and this can be difficult), drakes should not consume laying feed continuously during the laying season. At the conclusion of the laying season, ducks should be switched to a maintenance feed containing 0.60 to 1 percent calcium.

Breeder Rations

These feeds are formulated for birds that are producing eggs for hatching, rather than for eating. When compared with layer feeds, breeder rations need higher levels of some nutrients, especially vitamins A, B_{12}, and riboflavin. Other nutrients that normally are included at somewhat higher levels include vitamin E, pantothenic acid, pyridoxine, biotin, folacin, and the minerals copper, iron, manganese, and zinc. In general, calcium levels are about 0.5 percent lower in breeder rations when compared with layer rations. Breeder rations are designed to produce high fertility, strong embryos, and good hatchability. Game-bird breeder rations often work well with ducks — especially with strains that have a history of poor fertility and hatchability.

When breeder rations are not available, the following mix will usually give good results: eight and a half parts chicken layer, one part cat kibbles, and one-half part Animax or Calf-Manna (or similar product). Like layer feeds, breeder rations should be fed 2 to 3 weeks prior to and throughout the laying season and discontinued promptly when egg production ceases.

Complete Waterfowl, All-in-One Waterfowl

To keep things simple, some feed companies produce one waterfowl feed that they market as appropriate for all ages. Basically, these rations are starter/grower feeds that must be supplemented with calcium for laying birds — crushed oyster shells and ground limestone should be fed free choice during the laying season when these feeds are used. Because these feeds are essentially starter/grower rations, they can usually be supplemented with grains during some stages of a duck's life. For example, grains can be added at the following rates: 10 percent grains added from three to eight weeks of age and 20 to 25 percent from nine weeks until point of lay and for adult ducks during the nonbreeding/laying season.

Maintenance Rations

Feed manufacturers seldom sell poultry/waterfowl maintenance rations. However, game-bird maintenance feeds are available in some localities and normally work satisfactorily for ducks. Also, game-bird flight conditioner is used by some duck keepers with good results as a maintenance ration.

You can also make a maintenance ration by mixing 25 to 50 percent cereal grains with starter/grower feeds. How much grain should be added depends on the particular starter/grower feed used and on the environmental temperature. As a general rule of thumb, when temperatures are below freezing, 50 percent grains can be added, and when temperatures are above 80°F (27°C), no more than 25 percent grains should be added.

Concentrates

Manufactured concentrates such as Calf-Manna (Manna Pro) and Animax (Purina) are easily digested and have a high concentration of protein, vitamins, and minerals. These products can be useful during times when supplements are needed (the breeding season, for example) or when feed consumption is down due to old age, stress, injury, or high environmental temperatures. They are normally used at 5 to 10 percent of the ration. Due to their high concentration of nutrients, care must be taken to not feed excessive amounts, which could lead to toxic overdose.

Cat Kibbles

Because many modern poultry feeds are vegetarian, high-quality cat kibbles containing fish and/or meat are especially useful as a supplement for ducks during the breeding season, for ducklings, and for ducks of any age that are growing feathers. Animal protein helps birds grow superior feathers. Like any unfamiliar feed, ducks often will not eat them at first. Due to the relatively high fat content of kibbles, they are usually fed at no more than 10 to 15 percent of a duck ration for a prolonged period of time.

Vitamin and Mineral Supplements

Vitamin and mineral supplements are available in powder, liquid, capsule, and tablet form. Vitamin and electrolyte preparations that are formulated for poultry and game birds can be useful for getting newly hatched ducklings off to a good start and boosting the immune system of birds that are ill, recuperating from injuries, shipped, or stressed by old age or severe weather. Some vitamins and minerals are toxic when ingested in excessive quantities, so these supplements must be used with care.

Antibiotics

The proper use of antibiotics treats and reduces the occurrence of some diseases. Antibiotics work by killing bacteria in the body; unfortunately, they destroy both "good" and "bad" microorganisms. While they are useful for treating some diseases, their overuse has profound long-term negative consequences, including decreased efficacy of the drug, contaminated meat and eggs, and decreased natural resistance in a flock of birds. Always use antibiotics with care, and follow withdrawal recommendations carefully. Remember: Always follow recommended dosage and length of treatment — even if the bird appears to have recovered. Cutting short the treatment protocol can allow the organism to survive and build resistance to the antibiotic, which can result in superbugs.

Probiotics

Whereas antibiotics destroy microorganisms indiscriminately, probiotics are beneficial bacteria that actually boost the body's natural defenses against some diseases. In healthy animals the "good" microorganisms are able to keep the "bad" ones from multiplying sufficiently to endanger the health of the host.

Research indicates that supplementing the diet of animals with probiotics can result in healthier, more productive birds. Probiotics are especially helpful for getting newly hatched ducklings off to a good start and as a supplement to the diet of ducks that are old or otherwise stressed.

Insoluble Grit

To aid their gizzards in grinding food, ducks need to ingest insoluble grit such as crushed granite or coarse river sand. Feed stores sell granite grit in various sizes: The chick-size version should be used for young ducklings, and the hen-size for ducks six weeks and older. Cleaned river sand or small pea gravel can also be used.

Crushed Oyster Shells

Crushed oyster shells are commonly used to supplement the diet of birds with calcium. Because oyster shells are hard and dissolve relatively slowly, they also function as a grit. An advantage of oyster shells is that they provide a steady supply of calcium to the bloodstream 24 hours a day. This metering effect is especially important to laying ducks during warm and hot weather.

Growing ducklings and nonlaying adults should be fed oyster shells only if it is known that their diet is deficient in calcium. Oyster shells can give ducklings that are already receiving adequate calcium a harmful calcium overdose. Oyster shells are available in both chick and adult sizes.

Cereal Grains

Cereal grains make up the majority of most duck rations. However, by themselves they are not a nutritionally complete diet. Many nutritional deficiencies result from people assuming that ducks can thrive solely on a diet of grain. Worldwide, many kinds of grains are fed to ducks. Here, we will briefly look at the ones most commonly used in North America.

Corn is the most common ingredient in poultry feeds. It is high in energy, which makes it valuable as a cold-weather feed, but it does tend to make ducks fat. Corn is a rich source of linoleic acid (an essential fatty acid that can be deficient in wheat-based diets) and xanthrophylls (yellow pigments that enhance the color of skin, legs, bills, body fat, egg yolks, and feathers). Ducks are fond of corn and can overeat it. Corn is susceptible to molds that produce aflatoxins, which can be harmful or fatal to ducks. Never feed corn (or any grain) that looks or smells moldy.

Wheat is higher in protein and lower in energy than corn, and it is well liked by ducks. When hard wheat is fed prior to being sufficiently cured (45 to 90 days), some duck raisers have reported that their birds developed severe diarrhea. On the other hand, we have fed hundreds of tons of soft white wheat straight out of the field with no trouble. Ducks are fond of wheat and tend not to get as fat on it as on corn; however, wheat has approximately half as much linoleic acid as corn. A linoleic acid deficiency can result in poor growth, smaller egg size, and reduced hatchability. Birds fed wheat-based rations typically have pale-colored skin, bills, legs, and egg yolks.

Oats are lower in energy than corn and wheat but have more than four times the amount of fiber. Most ducks are not initially fond of whole or rolled oats and may take some time to learn to eat them. We include oats in our grower, breeder, and maintenance pellets, which eliminates any avoidance issues. In general, ducks that have 5 to 25 percent oats in their diet grow slower, are less inclined to feather eating, have fewer leg and wing deformities, have better-quality feathers, have less internal fat, produce more offspring, and live longer. Due to their high fiber content, oats normally are not included at more than 5 to 25 percent of the ration. Oats lack xanthrophyll pigments.

Barley, compared to corn, is higher in protein and fiber, has approximately 85 percent of the calories, is low in linoleic acid, and lacks xanthrophylls. Ducks are not particularly fond of barley, but its inclusion in the diet can reduce cannibalism and internal body fat. Young ducklings especially have some difficulty in digesting it, and it tends to make manure sticky. Whole barley is easier for birds of all ages to digest if it is soaked overnight in cool water at room temperature prior to feeding. We use no more than 10 percent barley in any of our pelleted rations.

Sorghum grain is readily eaten by ducks and is similar to corn in calories and protein level but significantly lower in linoleic acid and xanthrophylls.

Legumes and Cottonseed Meal

Legumes are the most frequently used protein supplement in poultry feeds today. However, due to the fact that substances they contain can inhibit digestion, be toxic, and suppress development and growth, they must be used with certain precautions. Raw legume seeds of any kind should not be fed to poultry under most circumstances, especially to growing and breeding birds.

Heat-treated soybean meal is used in most vegetarian poultry feeds as the main protein supplement. However, due to a trypsin inhibitor, *raw soybeans are not satisfactory* for use in monogastric animals.

Alfalfa meal (forage, not seed) is often used in duck feed as a source of protein, xanthrophylls, and various vitamins and minerals. Alfalfa does contain saponin, which restricts the amount that can safely be used in poultry feeds. Excessive amounts of alfalfa can cause tiny hemorrhages and increase the incidence of blood spots in freshly laid eggs. We limit the amount of alfalfa meal in our rations to 2 to 4 percent.

Other legumes need to be used with caution. Midway through one hatching season, we started experiencing decreased hatchability and the ducklings that did hatch had very sparse down. Eventually, we pinpointed the problem: the feed company — without telling us — had replaced a portion of the heat-treated soybean meal with raw peas. Once the ducks were fed the correct diet again for two weeks, the eggs they produced hatched normally. We have had several other reports of significant reproductive problems when breeder ducks were fed raw legumes.

Cottonseed meal is sometimes used as a protein supplement in poultry feeds. Because cottonseeds contain the toxic substance gossypol, cottonseed meal should be used with care. We do not use it at all, and I recommend limiting its use to no more than 2 to 4 percent of the ration.

Meat and Fish Meal Supplements

Traditionally, duck rations included fish and/or meat products, which mimicked their natural omnivore diet. Due to concerns of meat by-product–borne diseases, the strong trend today is for feed manufacturers to sell vegetarian diets for poultry. Under many circumstances, ducks perform better if they have at least 2.5 percent fish or animal products in their diet.

Texture of Feed

Feeds are available in a variety of textures. Each has advantages and disadvantages in how they are prepared and consumed.

Mash

In mash feeds, ingredients are ground up into coarse, medium, or fine flour consistency. Generally, grinding makes feeds easier to digest and promotes even mixing of the ingredients. Mash feeds are dry and powdery, "gum up" the bills of ducks, greatly increase feed waste, and are normally not recommended.

If fed to ducks, mash can be moistened (but not made soupy) with water or milk. When this method is used, only prepare a quantity that will be cleaned up in about an hour, and frequently clean feeders and mixing utensils. Moistened feeds can spoil rapidly and, if not carefully monitored, cause serious problems.

Pellets

Many feed companies run their mash through a pellet mill to make a firm pellet in order to reduce dustiness and waste. Pelleting is an additional expense in manufacturing feed, but the reduced loss of feed through spillage usually more than makes up for any price increase. Even though pellets come in a variety of sizes, most are too large for ducklings (especially smaller breeds) to consume for at least the first week or two.

Crumbles

Feed that has been pelletized can be crushed to form crumbles. The main advantage with crumbles is that birds of all ages can eat them. While there normally is more waste with crumbles than with pellets, small ducks can readily eat them.

Whole Grain

Once they are familiar with them, ducks will consume many types of whole grains. Ducklings, however, have difficulty eating all but the smallest grains. Due to a protective hardened outer covering, whole grains are more resistant to nutrient loss and spoilage than any other form. Whole grains are harder to digest (primarily a concern with very young or old birds or in situations where no insoluble grit is provided), and when they are used in mixed rations, ducks tend to pick out their favorites and leave the rest behind.

Cracked, Rolled, Crimped, Flaked Grains

When the protective outer coating of grain is broken, the nutrients are often easier to digest. However, the grain also becomes more susceptible to nutrient loss and spoilage. Therefore, freshness is critical.

16

Butchering

ONE OF THE MAIN REASONS DUCKS are raised is for their excellent meat. While they can be taken to a custom dressing plant to be butchered, this is an additional expense, and it robs you of the satisfaction of preparing your own food.

Cleanliness throughout the butchering process is essential to curb contamination and spoilage of meat. All cutting utensils should be sharp — dull knives are a waste of time as well as unsafe.

When to Butcher

One of the most time-consuming parts of the butchering process is the removal of the feathers. To make this job as simple as possible, butcher ducks when they are in full feather. If a duck is slaughtered when it is covered with pinfeathers, a picking job that normally takes 3 to 5 minutes can develop into a frustrating, feather-pulling marathon.

Depending on the breed, management, and season, ducklings are normally in full feather for only 5 to 10 days sometime between six and a half and ten weeks of age, except Muscovies, which require ten to sixteen weeks to feather out. Shortly after achieving full feather, young ducks go into a molt and begin replacing their juvenile garb with adult plumage. If ducklings are not dressed before this molt commences, they will not be in full feather again for approximately 6 to 10 weeks, when their adult plumage will be acquired.

Preparations for Butchering

Ducks should be taken off feed 8 to 10 hours prior to killing or the night before if they are going to be dressed early the following morning. To avoid excessive shrinkage, drinking water should be left in front of the birds until they are slaughtered.

Killing

The least enjoyable task in raising ducks is killing them on butchering day. But for those of us who raise our own meat, it's a necessary chore.

There are several ways of dispatching ducks. The simplest and most impersonal method is the ax and chopping block. It is advantageous to have a device on the chopping block to hold the bird's head securely in place (such as two large nails driven at a slight angle into a heavy block to form a V). A sharp cutting edge on the ax is a must. Instead of allowing the beheaded duck to thrash about on the ground, as soon as the head is removed, the bird should be hung by its legs or placed in a killing cone (see illustration below) to promote thorough bleeding and to prevent it from becoming bruised or soiled.

A second method of killing is to suspend the live duck head down with leg shackles or in a killing cone. Grasp the bill firmly with one hand; with the other hand stun with a blow to the back of the skull using an appropriate club (for example, a spare handle of a hardwood framing hammer). With a sharp knife cut the throat about an inch below the base of the bill, on the left side, severing the jugular vein, but leaving the head attached. The head provides a convenient handle if the duck is going to be scalded.

Killing cones and leg shackles allow the ducks to hang head downward during killing.

To avoid discolored meat in the dressed product, it is vital that ducks are bled thoroughly and not allowed to thrash around on the ground before they are processed further.

Plucking (Picking)

The sooner a duck is picked after it is bled, the more easily the feathers will come out.

Dry Picking

The best-quality feathers for filler material and the most attractive carcasses are obtained when ducks are dry-picked. Novices often find this method unbearably slow. Veterans, on the other hand, can dry-pick a duck clean in 3 minutes or less. One secret for success is to extract feathers in the same direction they grow and have extremely strong fingers! Pulling feathers against the grain invariably results in torn skin. For a better grip, lightly dust the duck with resin or periodically moisten your hands with water. The large plumes of the wings and tail need to be plucked out one or two at a time.

Because feathers and down float in every direction when waterfowl are dry-picked, choose a setting that is free from drafts. Picking into a large plastic garbage bag and covering the floor around the plucking area with a tarp or an old sheet can help keep feathers under control.

Scald Picking

The most common method for defeathering ducks is to scald them prior to picking. The only equipment needed is a large container in which to submerge an entire duck and a utensil (a forked branch or a 1 × 2 × 24-inch [2.5 × 5 × 60 cm] board with a large nail driven in at an angle near one end), which is used to hold the bird under water.

To scald, hold the duck by its feet and wedge the neck into the fork of the stick. Then dip the bird up and down in water that is 125 to 145°F (50 to 63°C), making certain that the water penetrates through to the skin. To improve the wetting ability of the scalding water, a small amount of liquid dishwashing detergent can be added.

The scalding time for ducks varies from 1 to 3 minutes, depending on the age of the bird and the temperature of the water. Mature ducks require longer and hotter scalds than ducklings. If you find that the feathers are still difficult to remove after the initial scald, the bird can be redipped. However, overscalding causes the skin to tear easily and discolors the carcass with dark blotches.

Ducks should be picked immediately after scalding, starting with the wing and tail feathers. Because the feathers are going to be hot, have a bucket of cold water nearby to dip your hands into as needed.

If your first attempts at scald-picking poultry do not produce carcasses that are as attractive as those that are processed commercially, do not get discouraged. Your results should improve once you've gained a little experience. Feathers from scalded ducks are of good quality when handled correctly (see Care of Feathers, page 264).

Wax Picking

A popular variation of the scalding method is to dip rough-picked ducks in hot wax. Even when birds with a moderate number of pinfeathers are butchered, this procedure can produce clean carcasses in a short time. Paraffin can be used, but a mixture of one part beeswax to one part paraffin is preferable. There are also products such as Dux-Wax that are made specifically for this purpose and are available from some poultry-supply distributors (see appendix G, page 339).

When wax is used, ducks are scalded without detergent in the water and are rough-picked by removing the large tail and wing feathers and approximately 90 percent of the body plumage. Prior to scalding, place a separate container of solidified wax in a large receptacle of hot water (double boiler style) where it is melted and heated to the desired temperature. You may want to experiment to find the wax temperature that works the best for your situation, but a good starting point is 145 to 170°F (63 to 77°F). The hotter the wax, the better it tends to adhere to the feathers and skin. Cooler temperatures produce a thicker coating that is easier to pull off. Ideally, there are two containers of wax: one at approximately 170°F for good penetration and one at 145°F to build up the wax.

CAUTION

Wax is extremely flammable and when heated can cause deep skin burns if it splashes on anyone nearby! Do not use wax around open flames and avoid direct contact with any heat sources. Extreme care must be taken when working with hot wax to avoid burns and fires. To protect your arms and hands, gloves may be used during the waxing process.

Rough-picked ducks are dipped into the hot wax several times. If they were scalded prior to picking, they will need to be partially dried (old clean towels work well) before dipping so that the wax will adhere to the feathers and pinfeathers. If the wax container is large and deep enough, the duck can be held by its legs for dipping; otherwise, the head can be held in one hand and the legs in the other hand to dip into a shallower container. After each dip in the wax, spray or dunk in cold water or wait long enough between each dunking to allow sufficient congealing that will build up a good layer of wax. If only a few birds are being dressed, you may find it simpler to melt a small container of wax to be poured over the carcasses.

After sufficient wax buildup, submerge the waxed duck in cold water long enough to harden the wax and grip the feathers. The wax and feathers are then stripped off together, resulting in a finished product that is clean and attractive. Birds that are extra pinny can be rewaxed.

Used wax can be recycled by melting it and skimming off the feathers and scum and boiling out any existing water.

Singeing

After ducks are picked, long, hairlike filament feathers usually remain. The simplest way to remove these filoplumes is to pass the carcasses quickly over a flame, being careful not to burn the skin. A jar lid with a thin layer of rubbing alcohol in the bottom gives the best flame I know of for singeing. Alcohol burns tall, cleanly, and odor free. Newspapers (do not use colored sheets) loosely rolled into a hand torch, gas burners, and candles can also be used.

CARE OF FEATHERS

Duck feathers are a valuable by-product of butchering. If you plan to save the feathers, keep the down and small body feathers separate from the large stiff plumes of the wings, tail, and body as the slaughtered birds are being picked.

When ducks are scalded prior to picking, the feathers need to be washed with a gentle detergent, rinsed thoroughly in warm water, and spread out several inches thick on a clean, dry surface or loosely placed in cloth sacks and hung in a warm room. Stir the wet feathers daily to fluff them and ensure rapid drying. Once they are well dried, feathers can be bagged and stored in a clean, dry location. (See appendix E, page 335, for instructions on how to use feathers.)

Skinning

Ducks can be skinned rather than picked. Some advantages of this technique are that ducks with pinfeathers can be dressed as easily as those in full feather and some people find skinning less time-consuming than picking. Because skin is composed largely of fat, skinning significantly reduces the fat content of the dressed duck.

The major drawbacks of this method are that skinned carcasses lose much of their eye appeal when roasted whole; special precautions must be taken in cooking the meat to prevent dryness; some flavor is lost; and a higher percentage of the bird is wasted.

To prepare a duck for skinning, remove its head and the last joint of each wing. The duck can then be hung by its legs (from strong shackles or cords) or the feet removed and the bird laid on its back on a table. Slip the blade of a small, sharp knife under the skin of the neck, and slit the skin the length of the body, cutting around both sides of the vent. The final step is to peel the skin off, which requires a good deal of pulling. Over stubborn areas a knife is needed to trim the skin loose.

Eviscerating

Ducks can be drawn immediately after they have been defeathered, or they can be chilled in ice water for several hours or hung in a cool (33 to 36°F [0.5 to 2°C]) location to ripen for 6 to 24 hours. Chilling the carcasses first has the advantage of making the cleaning procedure less messy, and aging before eviscerating produces stronger-flavored meat, which is preferred by some people.

With the bird resting on a clean, smooth surface (we use a decommissioned cookie sheet), remove the feet, neck, and oil gland, making certain that both yellowish lobes of the oil gland are cut out. Then make a shallow 3-inch- (7.5 cm) long horizontal incision between the end of the breastbone and the vent, being careful not to puncture the intestines that lie just under the skin.

Through the incision insert your hand into the body cavity and gently loosen the organs from the inside walls of the body, and pull them out. Cut around the vent to disconnect the intestines from the carcass. The gizzard, heart, and liver can be cut free and set aside before the unwanted innards are discarded.

The esophagus (ducks do not have true crops) and windpipe are well anchored in the neck and require a vigorous pull to remove them. The pink,

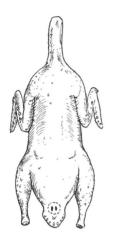

Cut off both lobes of the oil gland located on top of the tail.

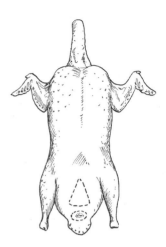

Make an incision between the end of the breastbone and the vent to remove innards.

spongy lungs are located against the back among the ribs and can be scraped out with the fingers if you desire.

To clean gizzards cut around their outside edge and then pull the two halves apart. The inner bag with its contents of feed and gravel can then be peeled away and discarded. The final step is to rinse the muscular organ with water.

The gallbladder, a small green sac, is tightly anchored to the liver and should be removed intact since it is filled with bitter bile. A portion of the liver must be cut off along with the gallbladder so bile does not spill onto edible meat.

Unwanted feathers and body parts make excellent fertilizer and should be buried near a tree or in the garden sufficiently deep that domestic or wild animals will not dig them up. (The exact depth depends on your soil type, but in our moderately heavy clay loam soil, we use a cover depth of at least 18 inches.) Do not feed raw innards to cats and dogs. If you give uncooked entrails to your pets, they can develop a taste for poultry and may kill birds to satisfy their cravings. Also, throwing offal across the fence is a sure way to draw predators from near and far.

Cooling the Meat

After all the organs have been removed, thoroughly wash the carcass and chill it to a temperature of 34 to 40°F (1 to 4.5°C) as soon as possible. If meat is cooled slowly, bacteria may multiply, causing spoilage and off flavor. Poultry can be chilled in ice water or air-cooled by hanging the carcasses in a refrigerator or in a room with a temperature of 30 to 40°F (−1 to 4.5°C).

Packaging and Storing Meat

After the body heat has dissipated from the carcasses, they should be sealed in airtight containers. If the meat was chilled in water, it should be allowed to drain for 10 to 20 minutes before it is packaged. To retain the highest quality in meat that is going to be frozen, suck the air from plastic bags with a straw or vacuum cleaner before sealing. To produce the tenderest meat, poultry must be aged 12 to 36 hours at 33 to 40°F (0.5 to 4.5°C) before it is eaten or frozen.

BUTCHERING CHECKLIST

- Remove the ducks' feed 8 to 10 hours before they are killed.
- Sharpen all cutting tools.
- When catching ducks prior to butchering, keep them calm and handle them carefully to avoid discolored meat due to bruises.
- Hang slaughtered ducks in killing cones or by their feet to avoid bruising or soiling the meat and to ensure thorough bleeding.
- Remove feathers as soon as possible after birds are bled.
- Singe off filament feathers.
- Trim out the oil gland from the base of the tail.
- Cut off feet and shanks at hock joints.
- Make a horizontal incision between the vent and the end of the breastbone, and gently lift out innards.
- Set aside the heart, gizzard, and liver before disposing of unwanted innards.
- Pull out the windpipe and esophagus from the neck/chest area.
- Extract the lungs from between the ribs.
- Rinse the carcass thoroughly with clean, cold water.
- Clean the gizzard, and carefully remove the gallbladder from the liver.
- Chill dressed duck to an internal temperature of 40°F (4.5°C).
- Bury unwanted body parts, entrails, and feathers in the garden or near a tree, sufficiently deep so that scavengers will not dig them up.
- Age the meat for 12 to 36 hours at 33 to 40°F (0.5 to 4.5°C) prior to cooking or freezing.
- Package and freeze the meat, or enjoy a festive banquet of roast duck.

17

Health and Physical Problems

COMPARED WITH CHICKENS, ducks have greater resistance to many diseases and parasites. Providing ducks with a proper diet, reasonably clean drinking water, adequate shelter, and sufficiently clean living quarters with ample space minimizes health problems. In my 50 years of raising thousands of ducks and every recognized breed, I have never had to treat a home-raised broadbill for coccidiosis or internal parasites or had to use a vaccine for a communicable disease.

The most common causes of health problems in ducks are improper diet, ingestion of toxic substances, overcrowding, filthy pens, and injuries. If you are unable to diagnose a malady, promptly seek advice from an experienced waterfowl breeder, veterinarian, or animal diagnostic laboratory. Procrastination may prove fatal and can result in an epidemic. Many infectious diseases cannot be accurately diagnosed without proper lab tests; check with your local veterinarian for the nearest diagnostic lab.

Sick or injured ducks should be penned in a dry, warm, clean enclosure, and provided a balanced diet and clean drinking water (if a contagious disease is suspected, the sick pen should be isolated, well away from other birds). Dead birds should always be removed as quickly as possible to avoid attracting predators and, for communicable diseases, to curb the likelihood of infecting healthy birds. Burn carcasses in an incinerator or bury them deep enough that they will not be dug up by scavengers. Leaving deceased birds lying around or tossing them over the fence into the bushes is an invitation to trouble.

Categories of Health and Physical Problems

Based on their origin, health and physical problems can be grouped into five main categories: nutrition and toxic substances, infectious organisms, injuries, parasites, and genetic defects.

Nutrition and Toxic Substances

The leading causes of health and physical problems in ducks are improper diet and ingestion (and rarely, inhalation) of toxins. Therefore, if ducks in your flock are experiencing health problems, but display no obvious signs of injury or infectious disease, the first step is to determine if diet is the culprit (I am using the term "diet" to include anything the birds ingest, including contaminated soil and water and poisonous plants and animals).

To determine if health problems are related to something the ducks are ingesting, you can (1) pen birds in a clean pen where you can be sure they are not consuming poisonous plants, invertebrates, baits or sprays, leaded paint, or contaminated soil or water; and (2) change their feed to a well-known brand that is balanced for all nutrients (in my experience Purina typically has high quality-control standards and balanced formulas).

The most common causes of health problems related to diet include (1) nutritional deficiencies, (2) nutritional overdoses (where vitamins and minerals are toxic in excessive amounts), (3) nutritional imbalances (the ratio of some nutrients to each other is as important as the quantity — phosphorus and calcium, for example), (4) medication toxicity, (5) molds, (6) spoiled feed, (7) toxic plants or animals, (8) pesticides (which often are harmful to ducks even if the manufacturer claims otherwise), (9) baits, (10) leaded paints or solvents, (11) contaminated water or soil, and (12) rotting plant or animal material.

A common mistake is to presume that a health problem cannot be diet related because only part of the flock is showing symptoms. Remember, every bird digests and assimilates nutrients and toxins differently and it is normal for diet-related symptoms to vary widely in a flock of birds, from individuals that show no symptoms to others that experience severe symptoms or death. Furthermore, some breeds, varieties, and even strains within a variety have, as a group, differing dietary needs and sensitivities. A diet that is adequate for one duck may have serious consequences for another.

On the next page are some examples of diet-induced health problems in ducks.

Salt Deficiency

A few years ago a number of waterfowl raisers along the East Coast reported that their young birds showed symptoms that included listlessness, irregular growth, poor feather quality, and, in some cases, increased mortality. It was then observed that everyone whose birds were affected used the same brand of grower feed. A feed analysis showed that several vitamins were marginally low, and the salt content was well below the minimum requirement. The feed company later acknowledged that salt had been inadvertently omitted.

Excessive Calcium

One year about 10 percent of our five- to eight-week-old ducks began looking listless, and small spots of blood were discovered in their droppings. I immediately suspected coccidiosis and rushed several birds to the veterinary diagnostic lab at Oregon State University. The lab report came back negative for coccidiosis or for any infectious disease. I called the mill that custom mixes our feed, and they assured me that the correct formula had been used.

After a second negative lab report, I switched the birds to a national brand of grower feed, and within a few days the symptoms gradually disappeared and there was no more mortality. After our feed company did further checking, they discovered they had accidentally doubled the calcium level from 1 to 2 percent in the last batch of our grower pellets. Therefore, the growing birds were suffering from a phosphorus-calcium imbalance, as well as excessive calcium.

Mycotoxin-Contaminated Feed

Mycotoxins are produced by molds. When excessively high levels are in their feeds, some ducks in a flock will not thrive and mortality rates will increase.

Vitamin Deficiency

A number of years ago, a company sold a feed that it advertised as being adequate for all species and ages of poultry. When this ration was fed to ducklings, some of them grew slowly and/or developed bowed legs. When an independent lab analyzed the feed, it found that for ducklings it contained borderline deficiencies of several vitamins, including niacin. Just because a ration is recommended by a company or dealer does not guarantee that it is adequate for ducks.

Infectious Organisms

When ducks are raised in relatively small flocks with plenty of room and a calm environment and consume a proper diet that supplies all essential nutrients in adequate quantities, they are impressively resistant to infectious diseases. Ducks that are stressed by crowding, filthy living conditions, or inadequate diet are more prone to succumb to infectious diseases.

Large commercial duck operations are more likely to encounter infectious diseases due to the high concentrations of birds. It sometimes becomes necessary for commercial operators to routinely medicate or vaccinate for highly infectious diseases.

Infectious diseases can be spread via wild birds and animals, infected ducks from other flocks, hatching eggs, clothes and footwear, and air or water. Some infectious diseases spread overnight while others may take weeks to move through a flock. Quick diagnosis and treatment are important for reducing severity and mortality.

Medications and antibiotics should always be used with care and only when necessary to minimize accidental toxicity or the development of resistant strains of disease organisms. When antibiotics are used it is important to complete the recommended treatment, even if the birds appear healed prior to completion. Some medications and antibiotics are toxic to some strains of ducks and not others; it is safest to administer medication to a few birds and look for negative reactions before treating an entire flock.

To find out if there are infectious diseases that are common in your area, consult with local duck raisers and veterinarians, Extension agents, and regional diagnostic laboratories.

Injuries

Injuries are kept to a minimum when ducks are protected from predators, kept in pens free of sharp objects and obstructions to trip over, provided a calm environment, and handled properly and at a minimum. Due to their anatomical structure, the hips, legs, and feet of ducks are the most susceptible to injury. People often ask if they should "put down" a bird that has suffered major injuries. In my experience birds appreciate the opportunity to recover. Ducks have an amazing ability to heal, even from injuries that appear to be catastrophic.

Parasites

Under normal circumstances, ducks have good to excellent resistance to most internal and external parasites. However, ducks that consume an inadequate diet or are crowded or forced to live in a filthy environment are more susceptible to parasites.

Genetic Defects

No matter how carefully breeding stock is selected and managed, some offspring will be produced with genetic physical defects. By definition purebred animals are at least mildly inbred (this homogamy is the reason individuals of the same breed share similar characteristics). Therefore, they are inclined to have more genetic defects than hybrids. Generally, the more intensely a strain is selected for specific characteristics, the more frequently genetic defects appear. When raising young stock, do not be surprised if you need to cull out some defective specimens.

Biosecurity

In this day and age of increased travel, microorganisms commonly hitchhike on human or wild bird and animal hosts from places far and near. Therefore, it becomes increasingly important to minimize the spread of disease organisms. As the cliché goes, "An ounce of prevention is worth a pound of cure." Some commercial operations have elaborate biosecurity measurements, but even small-flock owners can use the following strategies to greatly reduce the spread of disease organisms and the risk of dangerous infections among their ducks.

Within a farm, footwear should be changed or washed prior to entering the young-bird area, since ducklings are more susceptible to disease organisms than are older stock. Dead birds should be incinerated or deeply buried. Wild birds and rodents are potential disease and parasite carriers, so be sure to keep them out of buildings and feeders.

Because diseases are commonly spread on clothing and footwear, visitors to farms should not be allowed near or in the bird pens (unless they are provided clean overclothes and footwear). If you are visiting a bird farm, never assume you will be allowed into the pens.

Quarantining Birds

Birds that look healthy can be carriers of disease and parasites. Small-flock owners should treat new adult birds for internal and external parasites and quarantine them for 3 to 4 weeks, observing them for signs of disease prior to introducing them into the flock.

For further security commercial farms and small-flock owners should keep new arrivals isolated for a minimum of 90 days, since some diseases have a lengthy incubation period. For maximum security birds can be quarantined for a full year and their eggs hatched in separate facilities. This provides protection against diseases transmitted via eggs. After caring for quarantined birds, change or wash your footwear before returning to the main facility.

FIRST-AID KIT FOR DUCKS

- Antibiotic ointment/bacitracin for minor cuts and abrasions
- A medicated ophthalmic ointment for eye injuries and infections
- A broad-spectrum antibiotic (such as amoxicillin, tetracycline, or penicillin) to aid in the healing of infections
- A poultry or bird vitamin mix for adding to drinking water during times of stress
- A pyrethrin-based insecticide for treating external parasites and spraying around wounds to prevent maggots

---

Working with Your Veterinarian

A useful tool in maintaining a healthy duck flock is a good working relationship with a veterinarian. A veterinarian can prescribe medications, help diagnose problems, do emergency surgery and injury repair, and provide information regarding infectious organisms. There are a growing number of avian specialists in the veterinary field. To find one near you, contact the Association of Avian Veterinarians.

Diseases, Physical Disorders, and Parasites

Of the following ailments, those that are the most commonly seen disorders in small duck flocks are marked as such.

Aspergillosis *(Common)*

Aspergillosis is a disease affecting the lungs, primarily of young ducklings. It is indicated by gasping or labored breathing, poor appetite, and general weakness, and occasionally is accompanied by sticky eyes. The lungs of affected birds often contain small yellowish nodules about the size of a BB shot. Ducklings are most susceptible the first few days after hatching.

Causes. Commonly known as brooder pneumonia, it is the result of *Aspergillus fumigatus* mold being inhaled into the lungs when ducklings are hatched in contaminated incubators or from being brooded on moldy bedding or fed moldy feed.

Treatment. There is no known cure, but to prevent its spread infected birds should be removed, the brooding area and equipment disinfected, and the feed and bedding checked for musty odor or signs of mold.

Prevention. Hatch eggs only in thoroughly disinfected incubators and use mold-free bedding and feed.

Blackflies and Leucocytozoon Disease

Leucocytozoon disease is most common in Canada and subarctic zones and primarily affects young, unfeathered birds. Sudden death is the most common symptom in ducks. Lab tests (blood smears) are required for positive identification.

Cause. Blackflies (Simuliidae) are bloodsuckers and transmitters of leucocytozoon disease.

Treatment and Prevention. Prevention is the only effective treatment I am aware of in ducks. People in areas where blackflies are problematic have found that most losses can apparently be avoided if ducklings are well grown by the time blackflies emerge in the spring.

Botulism (Common)

Botulism, commonly known as limberneck, is usually fatal and can affect birds of any age. A few hours after eating poisoned food, birds may lose control of their leg, wing, and neck muscles. In some cases body feathers loosen and are easily extracted. Ducks that are swimming when paralysis of the neck develops often drown. Dying birds may slip into a coma several hours before expiring. Botulism normally kills in 3 to 24 hours, although in mild cases birds may recover in several days.

Causes. This deadly food poisoning is caused by a toxin produced by *Clostridium botulinum* bacteria, which are commonly found in soil, spoiled food, and decaying animal and plant matter. Botulism strikes most frequently in dry weather, when levels in ponds and lakes drop, leaving decaying plants and animals exposed for ducks to eat. Maggots that feed on decaying carcasses often carry the botulism toxin. Ducks can also contract this toxicant from spoiled feed or canned food from the pantry.

Treatment. All ducks suspected of having eaten poisoned food should be confined to a clean, shady yard or building and immediately provided fresh drinking water with a laxative added — either 1 pint (0.5 L) of molasses or 1 pound (0.5 kg) of Epsom salts per 5 gallons (19 L) water. Birds that cannot drink on their own should be treated individually. The addition of one part potassium to three thousand parts drinking water or individual doses of 1 teaspoon castor oil have also been recommended as treatments. In birds that are particularly valuable, flush out the contents of the esophagus with warm water by using a funnel and rubber tube inserted into the mouth and several inches down the esophagus. To avoid further problems every effort must be made to locate the source of botulism. A vaccine has been developed, but it is rather expensive and often difficult to obtain on short notice.

Prevention. Bury or burn carcasses of dead animals and clean up rotting vegetation. Do not let your ducks feed in stagnant bodies of water or give the birds spoiled canned goods or feed.

Broken Bones

The bones of birds have a wonderful ability to mend themselves. However, to prevent a duck from being permanently disfigured or crippled, it is often helpful to set and immobilize a wing or leg that is fractured.

Setting. Broken bones should be treated promptly, within 24 hours after the accident. A bone is set by gently pulling apart and, if necessary, slightly twisting the two halves until they mesh properly.

Splints. Broken bones should be held in alignment with splints (popsicle or tongue sticks often work well). A rigid support should be positioned on either side as far above and below the fracture as possible, and held securely in place with strong tape. The patient should be checked frequently to ensure that the brace is staying in place and blood circulation is not being restricted. Splints can normally be removed in 14 to 28 days.

Broodiness (Common)

Broodiness normally is not a sign of illness, although duck-raising novices are often alarmed by the behavior of broody females. The duck stays on her nest for long periods; quacks loudly when disturbed or when off the nest to eat, drink, and bathe; becomes protective of the nest; retracts her neck and head down onto her shoulders; and defecates a large amount of foul-smelling excrement soon after leaving the nest or while on it if startled.

Cause. Broodiness is caused by physiological changes in female birds that cause an increase in body temperature and give her a strong urge to set on and protect a nest or brood babies.

Setting females are susceptible to external parasites such as mites, lice, and stinging ants, and she and her nest should be treated with an appropriate insecticide if needed. Females allowed to set on a nest well past the normal incubation period are susceptible to nutritional deficiencies and may waste away and die if not removed to a well-lit pen with no nests and provided a balanced diet and bathing water.

Choking (Common)

Ducks will occasionally get feed caught in their throats. Normally, after a vigorous shaking of the head, the passageway is cleared and breathing returns to normal. However, sometimes a bird is unable to clear its throat and will suffocate if not promptly aided. When a duck obviously needs your assistance, pull its head forward until it is in a straight line with the neck, open the bill by squeezing with the thumb and index finger on both corners of the bill, and push your finger, a piece of ¼-inch (0.65 cm) rubber tubing, or the eraser end of a new pencil down the bird's throat until the obstruction is dislodged.

Chronic Respiratory Disease (CRD, Mycoplasmosis)

Chronic Respiratory Disease is an infection of the respiratory system and is most common in times of stress (see also Sinus Infection). Symptoms include coughing, sneezing, thick mucous discharge from the nostrils, ruffled feathers, and irregular or stunted growth. Instead of infecting all birds at once, CRD tends to move through a flock rather methodically over the course of several days to weeks.

Cause. CRD is caused by the highly contagious microorganism *Mycoplasma gallisepticum* (MG). CRD outbreaks are most common after birds have undergone stress due to being shipped, moving to new pens, sudden feed changes, onset of egg production, drastic weather changes, diet, and so on.

Treatment. At the first sign of diagnostic symptoms, administering an appropriate antibiotic can provide relief. Most strains of MG are sensitive to various antibiotics such as erythromycin (sold as Gallimycin) and tetracyclines (Aureomycin and Terramycin). Tylosin (Tylan) is often the most effective, although it can be hard to locate (ask your veterinarian).

Prevention. The best ways to prevent CRD are to acquire mycoplasma-free birds and provide an adequate environment and proper care. Unfortunately, birds that have had CRD can be carriers even after symptoms have long disappeared. Female carriers can pass mycoplasma to their offspring via eggs. Pens and cages that have housed infected birds should be thoroughly disinfected prior to reuse.

Chronic Wet Feathers

Normally, ducks in good feather condition are able to stay relatively dry even in prolonged periods of wet weather. Birds with chronic wet feathers look waterlogged during wet weather or after bathing.

Causes. There are a variety of causes, including a plugged or infected oil gland; heavy infestations of mites (some kinds are difficult to see); feathers covered with fuel, sprays, or other foreign substances; and living in sloppy pens.

Treatment. If the oil gland is infected or plugged, it should be massaged gently several times daily with a warm compress, and the bird should be given an oral antibiotic as prescribed by your veterinarian.

For mites an appropriate insecticide should be applied. In difficult cases some people have reported success with 1 percent injectable ivermectin for cattle and swine (*not* ivermectin plus clorsulon). Squirt or drip 0.10 cc per 4 pounds (1.75 kg) of bird down the patient's throat. (Because the active ingredient of this systemic vermicide is not registered for use in birds, ivermectin should be used only under the supervision of a veterinarian.) Birds should be treated again for mites in 4 weeks. Due to damaged feathers, ducks sometimes have to molt old feathers and grow a new set before their feathers shed water normally again.

Prevention. Keep ducks in a clean environment. Do not allow them to bathe in water that is contaminated with foreign substances, and treat them for external parasites as needed.

Coccidiosis

Coccidiosis is a major poultry disease, but unless invited by poor sanitation, it is seldom a problem in ducks. Symptoms include reduced appetite, ruffled feathers, heads drawn close to the body, and sometimes diarrhea and bloody droppings. In chronic cases birds may grow slowly and never attain full size or production or may waste away and finally die. In severe outbreaks large numbers of ducklings may die within a week or less. Because symptoms depend on the species of coccidia, it is recommended that birds be taken to a diagnostic lab to confirm coccidiosis outbreaks.

Cause. Coccidia are one-celled parasites that attack and destroy cells in portions of the digestive tract. There are numerous known species. These microscopic protozoa are generally present in moderate numbers where birds are raised, and they can survive in soil for more than a year. The egglike oocyst produced by coccidia can be transported on the shoes or clothing of people, in the droppings of wild birds, and by purchasing infected fowl.

Ducks generally have greater resistance to coccidiosis than chickens. However, when ducklings are overcrowded, brooded on damp litter, or kept in filthy quarters, they can suffer serious infestations. As birds mature, they normally develop immunity to "cocci."

Treatment. In general coccidiostats manufactured for chickens and turkeys will also be effective for ducks (one of the safest for ducks is amprolium, often sold under the names Amprol and Corid). These preparations can be added to feed or drinking water and are usually available from feed stores and poultry supply dealers. However, the recommended dosages for chickens and turkeys should be reduced by approximately one-third to one-half for ducks, since waterfowl consume greater quantities of feed and water and overdoses of some coccidiostats can be deadly.

Prevention. The best prevention is dry, clean bedding that is turned or changed frequently to promote dryness or, better yet, wire floors in the brooding area, or at least under and around the watering containers.

Cuts and Wounds

Compared with most other animals, ducks have a normal body temperature that is feverishly high (104–109°F [40–43°C]), which protects against some infections. In reasonably sanitary surroundings superficial scratches and abrasions usually heal naturally. However, when a duck sustains an open wound or is mauled, clinical care is needed. Prior to and after working on a wound, wash your hands thoroughly for a minimum of 2 to 3 minutes with warm, soapy water.

Treating Open Wounds

Ducks with deep or jagged cuts should usually have their feathers trimmed away from the wound's edges. Always hold a clean piece of gauze or lintless cloth over the wound while trimming feathers to prevent bits of webbing from adhering to the exposed flesh. Wash the wound with warm water and a mild soap; then rinse thoroughly with clear, warm water. Small pieces of shredded, loose skin that will not heal can be trimmed away.

To speed healing and prevent infection, apply a medicated ointment (such as Neosporin) once or twice daily. To keep flies away and prevent maggots, spray a pyrethrin-based insecticide on the feathers around the wound. If open cuts are not properly cared for, infections and maggots can be problems, especially in warm weather. If the bird has multiple or severe wounds, it can help to administer a broad-spectrum antibiotic, such as Eurofloxacin Baytril (available only by prescription), orally twice a day.

Sewing Up Gaping Wounds

Stitches are required when large patches of skin have been torn loose or deep lacerations sustained. While suture needles and silk thread are preferred, a sterilized sewing needle and white thread work satisfactorily for surface suturing. Each suture or stitch should be well anchored in the skin, but not over ⅛ inch (0.3 cm) deep. Sutures should be spaced approximately ⅜ inch (1 cm) apart and drawn tight enough to bring the two edges of the torn flesh together without much puckering. If nonabsorbing thread is used, stitches should be snipped and pulled out with tweezers in 4 to 5 days.

Distended Abdomen

It is normal for the abdomens of females to swell noticeably prior to and during the laying season. However, sometimes the abdomen, of either sex, can enlarge to the point that a duck has difficulty walking.

Causes. There are three main causes. First, ducks that are obese (most commonly in Heavyweight breeds) can accumulate large amounts of fat, and their abdomens will be rather soft when palpated. Second, due to oviduct malfunctions, either yolks or fully formed eggs are dumped into the body cavity, filling it with a firm mass. Third, due to the malfunction of a major organ, the abdomen can fill with fluid, in which case the paunch is extremely heavy and tight to the touch. Various factors make birds susceptible to "water belly," including excessive amounts of calcium in grower feeds, high-protein diets, ingestion of feed containing aflatoxins, and some disease organisms.

Treatment. Birds that are obese should be put on a lower-fat diet. For internal layers the only treatment is surgical removal of the egg mass. For "water belly" there normally is no cure. However, with special birds symptoms can often be temporarily relieved by a high-quality diet that is aflatoxin free and is balanced for all nutrients (with no more than 1 percent calcium), as well as the removal of the liquid by carefully inserting a large-gauge needle into the abdomen. However, because of the danger of puncturing internal organs, this procedure is best left to a veterinarian.

Prevention. The incidence of distended abdomen can be reduced by not letting laying birds become overly fat, providing the correct diet for birds at all stages of their lives, and providing aflatoxin-free feeds.

Duck Plague

Also known as duck virus enteritis, duck plague is a deadly disease that is quite contagious and can infect wild and domestic ducks, geese, and swans but not chickens or turkeys. Symptoms may present as sudden death of birds that are in good flesh, listless birds that are reluctant to move, extreme thirst, diarrhea (sometimes bloody), matted feathers on the head due to increased secretions from the eyes and nostrils, and protruding penis in dead adult males. High mortality in ducks of all ages is possible. Specimens must be sent to a diagnostic lab for positive diagnosis, and the disease must be reported to the state veterinary office.

Causes. Infections occur from direct contact with wild or domestic carriers or from water that has been contaminated by carriers of the highly contagious herpes virus. Waterways frequented by wild waterfowl that flow into pens can transmit the virus to domestic birds.

Treatment and Prevention. There is no known treatment, only eradication by depopulating the premises and thoroughly cleaning and disinfecting all possibly contaminated areas. Never introduce new birds that are carriers. In areas where duck plague is known to occur in wild waterfowl, domestic ducks should be isolated and kept off waterways frequented by wild birds. In regions where duck plague is known to occur, ducks that are kept on dry land are much less likely to contract this disease. A vaccine is available and is used by some commercial farms in susceptible regions.

Duck Virus Hepatitis

Duck virus hepatitis (DVH) is an acute, highly infectious disease primarily of ducklings under eight weeks of age. It is normally limited to commercial farms with large numbers of ducks in high concentrations. The disease spreads rapidly through an infected flock, with the first deaths often occurring within 24 hours to 3 days after exposure to the virus and an hour or less after the first symptoms appear. Mortality may be close to 100 percent in ducklings under four weeks of age.

Symptoms may include dull, listless birds that stop eating; watery, green diarrhea; purple-colored bills in white birds; and ducklings that retract their heads over their backs, fall on their sides, and paddle their feet. Some infected ducklings may recover rapidly, but they can continue to be carriers.

Cause. DVH is caused by a virus that can survive on contaminated equipment and in bedding for weeks. This disease is much more likely to occur when ducklings of different ages are kept in the same building (especially if they are moved from pen to pen), and when footwear, clothing, and vehicles are not disinfected when going between flocks. Also, birds such as starlings and English sparrows, as well as rats, can transport the disease from one flock to another.

Treatment. At the first sign of symptoms, antibody therapy can reduce losses. (This is normally practical only for large operations with ready access to the antibodies.)

Prevention. Good sanitation and keeping wild birds and rodents out of brooding and growing pens are keys to preventing DVH. This disease is much less likely on farms where each age group of ducklings is raised in a separate building that is thoroughly cleaned and disinfected after the birds are removed and prior to installing a new batch of hatchlings. Some commercial farms vaccinate their breeding stock for DVH, causing the ducks to produce antibodies that are passed through the eggs to ducklings. Ducklings can also be vaccinated.

Erysipelas

Erysipelas is relatively common in hogs and sheep, somewhat common in turkeys, and less common in geese. It is not common in ducks, but it is more likely if the ducks are raised in the same place as hogs, sheep, or turkeys with erysipelas.

Cause. Erysipelas is caused by the bacterium *Erysipelothrix rhusiopathiae*, which can live in soil for a long time. If ducks are exposed to it when they are under stress, high levels of sudden death can occur. Normally, erysipelas needs to be confirmed by a diagnostic lab to distinguish it from other causes of sudden death. Because erysipelas can cause a skin disease in humans, if you suspect that you are dealing with this disease, handle sick and dead animals with care, and use protective clothing and gloves.

Treatment. Erysipelas needs to be treated under the supervision of a veterinarian. Recommended treatment for all the birds in a flock where some birds show symptoms includes subcutaneous or intramuscular injection of procaine penicillin along with the appropriate vaccination. Ducks showing symptoms should be penned separately from the flock and given, in addition to the above, an intramuscular injection of potassium penicillin.

Prevention. The best prevention is good sanitation and management and keeping ducks off ground known to have had infected hogs, sheep, turkeys, or geese. In areas where erysipelas is a persistent problem, birds can be vaccinated with a bacterin (consult your veterinarian for details).

Eye Injuries, Foamy Eye *(Common)*

Eye problems can be a result of either injuries or disease processes.

Causes. Injuries often cause a bubbly foam to cover part or all of the eye. They most often occur during fighting and mating but can also occur routinely. The incidence of foamy eye increases during times of stress and sudden weather changes.

Unfortunately, foamy eyes can also be a symptom of infectious diseases such as Chronic Respiratory Disease. However, if foamy eye appears overnight and is not accompanied by unusual nasal discharge or other respiratory symptoms such as coughing and wheezing, then it probably is the result of an injury or irritant.

Treatment. Usually, time and protection from further injury are all that is required. Twice-daily application of an antibiotic ophthalmic ointment to an injured eye sometimes speeds healing. Some people have reported good results using the NFZ Puffer. The applicator is shaken and the powdered nitrofurazone is "puffed" into the bird's eye. To prevent accidentally inhaling or getting the powder in your own eyes, and to ensure that the medication gets in the patient's eye, only use the puffer in a protected area out of the wind. Use in an area where drifting medication cannot contaminate drinking water.

Prevention. Especially during the mating season, some drakes persistently fight with each other and should be separated to prevent serious injury. An excessive number of drakes can greatly increase the risk of eye injuries to ducks, so the correct ratio of males to females is important. Sharp protrusions, such as projecting fencing wire and brambles, should be eliminated.

Feather Eating *(Common)*

Cannibalism of this type, where birds pull out and eat one another's feathers, is most prevalent among ducklings (especially Muscovies, Mallards, and wild species) that are brooded artificially.

Causes. Feather eating is usually the result of boredom but can also be triggered by excessively high brooding temperature, intense light, overcrowding, an unbalanced diet, or the lack of green feed.

Treatment and Prevention. At the first sign of feather eating, check brooder temperature, reduce light intensity (using blue or red bulbs often helps), and provide ducklings with sufficient space, a balanced diet, and adequate quantities of feed. As long as ducklings are kept inside, put tender green foods (cut in bite-sized pieces) such as grass, lettuce, and dandelions in front of them as much as possible. As soon as it is safe, allow ducklings access to grassy pens during the day.

Foot Problems *(Common)*

Corns and calluses often develop on the bottoms of a duck's feet and normally are not a problem. However, in some cases they become infected or develop deep, bleeding cracks, causing the bird to go lame.

Causes. Foot trouble can result from bruises, cuts, splinters, or thorns in the footpad or from ducks spending much time on dry, hard, or abrasive surfaces. However, a frequently overlooked cause is a dietary deficiency in biotin, pantothenic acid, riboflavin, or another vitamin or mineral necessary for healthy tissues. Bacterial infections (bumble foot) may also cause footpad or joint infections, with staphylococcal bacteria commonly involved.

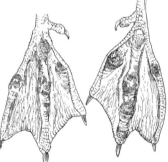

Bleeding cracks on the bottoms of feet may be a result of a diet deficient in biotin, pantothenic acid, or riboflavin.

Treatment. If a deficient diet is suspected, supplement the rations with a vitamin premix or feedstuffs (such as dried brewer's yeast, whey, dried skim milk, or alfalfa meal) that are rich in vitamin A, biotin, pantothenic acid, and riboflavin.

If the ball of the foot is inflamed, wash the foot with warm, soapy water; disinfect with rubbing alcohol; then remove splinter, if present. If you are sure there is a pus core, open the pad with a sharp, sterilized instrument (an X-Acto knife with a new blade or a single-edged razor blade can work here). Remove any pus or hard yellow core, disinfect the incision with iodine, and apply a medicated salve such as Neosporin.

Place the patient in a clean pen bedded daily with a layer of fresh straw (never sawdust or wood shavings) or clean towels, and provide a feed balanced for all nutrients as well as a small container of clean bathing water with 5 drops of chlorine bleach per gallon of water. Daily washing, disinfecting, and applying medicated ointment to the wound facilitate healing. For extra-valuable birds a dose of penicillin in tablet form for a minimum of 10 to 14 days seems to be helpful. In stubborn cases a bacteriologic culture to isolate the causative bacteria and an antibiotic sensitivity test may be done by a laboratory.

Prevention. Ducks that receive a balanced diet and that do not spend a significant amount of time on hard, abrasive surfaces seldom develop significant foot problems. Access to pasture and bathing water can reduce its occurrence.

Fowl Cholera

Fowl cholera is a highly contagious disease of both wild and domestic birds. In waterfowl, cholera often gives little or no warning, with apparently healthy birds dying suddenly. Chronic cases may be signaled by listlessness, lameness, swollen joints, diarrhea, breathing difficulties, and increased water consumption.

Waterfowl that die of an acute attack typically show little — if any — sign of the disease upon postmortem examination. In less severe cases the liver is often streaked with light-colored areas and spotted with minute hemorrhages and gray spots of dead tissue. It is fairly common to have tiny red hemorrhages on the intestines, gizzard, and heart that are visible to the naked eye. Also, the spleen may be enlarged.

Causes. Fowl cholera is caused by the bacterium *Pasteurella multocida*, which can survive in soil and decaying carcasses for several months or longer. It can be spread by wild birds, rodents, and scavengers or from ducks pecking at infected dead birds. Although cholera can occur any time of year, it thrives best in a damp, cool environment.

Treatment. Recommended treatment is one of the following sulfa drugs: sulfaquinoxaline sodium at the rate of 0.04 percent in drinking water, or 0.1 percent in feed for 2 or 3 days; sulfamethazine at 0.4 percent in feed for 3 to 5 days or sulfamethazine sodium, 12.5 percent solution at 30 mL per gallon of drinking water; or sulfamerazine sodium at 0.5 percent in feed for 5 to 7 days. Sulfa drugs must be used with caution, particularly with breeding stock, as they can be toxic. High levels of antibiotics such as tetracycline are sometimes used in the feed or injected under the skin. For the small-flock owner, the most practical treatment is usually adding easy-to-use pre-pared medications to the drinking water, such as Salsbury Sulquin. For sulfa-resistant infections, penicillin given intramuscularly is often effective.

Prevention. Sound sanitation practices are the best prevention. Water containers should be placed over wire-covered platforms and waterers frequently cleaned and occasionally disinfected with sodium hypochlorite (common bleach used at 4 ounces (0.1 L) per gallon (3.75 L) of water as a disinfectant) or an approved livestock sanitizer. Eliminate stagnant mud holes in the duck yard and burn or deeply bury all carcasses of dead birds and animals. In localities with a history of cholera, pasture rotation and vaccination using commercially available bacterins according to the manufacturer's recommendations may be necessary.

Frostbite

During freezing weather look for birds that have their feet frozen to the ground or ice. Also watch for ducks limping when forced to walk; swollen or red feet that feel hot to the touch; and tissue that is sloughing off. Unfortunately, frostbite is often not detected until lameness, gangrene, or discoloration occurs.

Cause. Prolonged exposure of the feet (or bare facial skin of Muscovies) to extreme cold can result in freezing of tissues.

Treatment. When waterfowl are found with their feet frozen to ice or the ground, pour warm water that is 90 to 105°F (32 to 40°C) — no hotter — over the frozen parts until they are freed. Then rapidly warm the frostbitten feet in a water bath (105 to 108°F [40 to 42°C]) for 15 to 20 minutes, and give the patient lukewarm drinking water. Do not rub the affected parts. If gangrene sets in, the frozen areas may eventually drop off or may need to be amputated and treated as an open wound. The oral administration of antibiotics such as penicillin and Terramycin to birds with severe frostbite reduces the chance of infection.

Prevention. Unless waterfowl have access to a large body of open water, ducks of all breeds should be enclosed in a yard or shed with a thick layer of bedding and provided protection from wind when temperatures fall below 20°F (−7°C).

Esophagus Impaction and Hardware Disease (Common)

Wire, nails, strings, and other objects may seem small and innocuous, but they can wreak havoc if ingested by fowl and can result either in esophagus impaction or in hardware disease. With an esophagus impaction, birds may slowly lose weight, stop eating, or sit around with eyes partially closed, apparently in severe pain. The obstruction is often visible as a lump, usually in the lower neck. With hardware disease, the symptoms are often similar to an esophagus impaction except that there will be no discernible lump or mass in the neck. Frequently, the foreign object will penetrate the muscle wall of the gizzard, causing peritonitis. A radiograph is the only way to confirm hardware disease in a live bird. When a postmortem is performed, the hardware is often found lodged in the gizzard.

Causes. Nails, bits of wire, pieces of string, blades of tough grass, excessive quantities of gravel, or other hard-to-digest objects are sometimes swallowed by ducks. When ingested, these objects may block or puncture some portion of the upper digestive tract.

Treatment. For an esophagus impaction the wad of material can sometimes be kneaded loose by gently massaging the compaction from the outside, working it back up and out of the mouth (tubing the bird with warm water, then holding it upside down while massaging the compaction is often helpful). If relief cannot be achieved by external methods, the blockage may have to be removed surgically. Having a veterinarian remove the impaction is preferable, but if you decide to do the surgery yourself, here are basic instructions for surgically removing an impaction.

Before starting the process, have an assistant firmly but gently hold the bird so that it is immobilized and you will have both hands free to do the surgery. Choose a clean, well-lit working area. For this operation, the bird does not need to be anesthetized.

To prepare for this operation, pluck the feathers directly over the impaction a few at a time until an area approximately 1½ inches (3.75 cm) in diameter is exposed. After washing your hands with soap and warm water for several minutes, drench the plucked patch with peroxide, gently tighten the skin by stretching it between the thumb and index finger, and make a shallow, inch-long (2.5 cm) incision through the skin with a sterilized (boiled for 3 minutes) knife, such as an X-Acto knife with a new blade. A second incision is made through the wall of the esophagus.

Using your finger or a sterilized, blunt instrument, remove the troublesome material from the esophagus and rinse with clean, warm water. Using a fine needle and gut suture material (gut must be used so that it will dissolve), draw the incised edges of the esophagus together with three or four single stitches that are tied off separately. The outer cut can be sewn in a similar manner, except that silk thread should be used.

When finished, wash the incision with peroxide, apply an antibiotic ointment, force-feed several capsules of cod-liver oil, and provide drinking water, but no feed, for 24 hours. Thereafter, supply small quantities of easily digested greens, such as lettuce, and an appropriate pelleted or crumbled feed several times daily until the stitches are removed. Apply an antibiotic ointment daily until signs of inflammation have dissipated. The outer sutures can be taken out after about a week.

For hardware disease, major surgery is usually the only remedy. If you deem the bird sufficiently valuable and can locate a veterinarian willing to perform this highly invasive surgery, it is possible, but risky, to remove the hardware from the gizzard.

Prevention. Never leave nails, wire, or string where birds can reach them. When hardware is used in an area to which birds have access, a strong magnet comes in handy for picking up dropped wire and nails.

It is inadvisable to place large quantities of grit in troughs with feed. If ducks have been without grit for some time, give only 1 teaspoonful per bird every other day for a week before giving grit free choice. I have seen situations in which it appeared that ducks that had been deprived of grit ate such large quantities of sand or pea-size gravel once it was available that their gizzards became impacted and the birds starved to death because feed could not pass through to the lower intestines.

Lameness *(Common)*

Lameness can have many causes, including dislocated hip, sprains of leg or foot joints, infections (cuts, abrasions, splinters), muscle damage, pinched nerves, dietary deficiencies (especially of niacin, biotin, and other B vitamins), or a calcium to phosphorus imbalance (common when high-calcium laying rations are fed to immature birds). Also important, but less common, are inherited leg weaknesses. Injuries can happen when ducks are chased, mauled by predators, handled improperly, startled at night (causing them to panic and run blindly into objects and walls), stepped on by larger animals, punctured by brambles and other sharp objects, or bruised by extended time on hard or rough surfaces.

Symptoms. Lameness can have sudden onset or develop gradually over days. Sprains are normally accompanied by swelling in the injured joint. Infections can also cause inflammation in the joints as well as other parts of the legs or feet. Pinched nerves can cause partial or complete paralysis of the leg. Leg problems associated with dietary deficiencies, calcium to phosphorus imbalance, or inherited weaknesses are usually signaled by legs that tremble when the duck stands still, give out after the bird walks or runs a relatively short distance, are bowed, or are twisted out at the hock joint (see also Spraddled Legs, page 298).

Treatment. If otherwise healthy, ducks with injured legs will normally recover if kept quiet in a clean pen and provided easy access to food that supplies a balanced diet and drinking water to which a good poultry vitamin mix has been added. Clean swimming water can help a duck with a serious leg injury to recover by allowing low-impact exercise.

When a deficient diet is the cause, take prompt action. Mixing a vitamin/ mineral supplement into the drinking water or with the feed (per manufacturer's recommendation) or the feeding of 2 to 3 cups (0.5 to 0.75 L) of dried brewer's yeast per 10 pounds (4.5 kg) of feed will often correct the problem. If you have been feeding immature birds a high-calcium laying ration, immediately switch to a feed that contains a maximum of 1 percent calcium and a phosphorus to calcium ratio of 1:1 to 1.0:1.5.

Prevention. Select breeding birds that display strong limbs from hatching to maturity. Don't catch or carry ducks of any age by their legs or run them across rough ground or over equipment such as water and feed troughs. Feed ducklings a diet fortified with niacin and vitamins D and A, and with the proper phosphorus to calcium ratio, in the range of 1:1 to 1.0:1.5. Young birds and nonlaying birds should not be fed laying rations, as they have too much calcium and an improper phosphorus to calcium ratio, which can cause serious problems.

Maggots *(Common)*

When open wounds are left untreated, especially during warm weather, blowflies may be attracted to the wound sites and may then lay eggs on sores. In a short time the eggs hatch into maggots and feed on surrounding tissue. During the breeding season the backs of female ducks can be lacerated as the drakes tread them. These lacerations can be prime sites for maggots.

Treatment. Prevention is the best treatment. Birds with wounds should have an antibiotic ointment applied daily until healed. As a precaution some people give wounded ducks an oral dose of ⅒ cc of 1 percent injectable iver-mectin for cattle and swine (*not* ivermectin plus clorsulon) per 4 pounds (1.75 kg) of bird.

Ivermectin gives protection against blowflies for approximately 2 weeks (a word of caution: ivermectin is not registered for use in birds and there-fore should be used only under the supervision of a veterinarian). In situa-tions where maggots are present, one technique is to spray a small amount of car-starting fluid onto the maggots. Some will fall off, and the others can be removed with a tweezers. Hydrogen peroxide can then be squirted into crevices with an eyedropper, which will cause embedded maggots to back out, facilitating removal. The wound should be treated daily with an antiseptic spray or ointment and checked for missed maggots until completely healed.

Prevention. Ducks that develop phallus prostration, prolapsed oviducts, or open wounds should be cared for promptly or destroyed to prevent a maggot infestation. Birds with wounds or open sores should have a pyrethrin-based insecticide sprayed daily on the feathers around the wound (cover the wound with a clean cloth to protect from the spray).

Niacin Deficiency (Common)

One of the most frequent problems that our farm gets calls for concerns young ducks suffering from niacin deficiency. Birds develop weak or bowed legs and often show stunted growth and enlarged hocks. In mild cases a clas-sic symptom is that some of the birds are of normal size and have strong legs while others are undersized and show leg deformities. (Rickets, sometimes confused with niacin-deficiency symptoms, is caused by a vitamin D_3 defi-ciency or a calcium or phosphorus deficiency or imbalance.)

Cause. This deficiency is caused by ducklings consuming a diet low in usable niacin. Most chick starters are deficient in niacin, but broiler starters typically have an adequate amount of niacin for ducklings. It is also important to remember that most of the niacin in plants is not available to ducklings.

Treatment. Ducklings exhibiting symptoms of a niacin deficiency can often be cured by the addition of 100 to 150 mg of niacin per gallon (3.75 L) of drinking water until they are eight to ten weeks old (niacin in tablets or powder is available at drugstores). If niacin deficiencies are not treated in time, ducklings can be permanently stunted or crippled.

Prevention. Ducklings must be fed a diet providing 35 mg of available nia-cin per pound (0.5 kg) of feed from zero to two weeks of age and 30 mg from two to ten weeks of age. If regular chick starter is used, add niacin in the drinking water at the rate of 100 to 150 mg per gallon from zero to ten weeks of age. When allowed to forage in areas with an abundant supply of insects, ducklings seldom experience a niacin deficiency.

Nutritional Deficiencies (Common)

There are many possible symptoms; the most common ones in ducklings include wide variation in body size in flocks with birds of the same age, stunted growth, retarded feather development, weak or deformed legs, and birds with reduced resistance to disease or parasites. Adult ducks that are undernourished produce poorly, often have rough-looking feathers, may be thin, and are susceptible to disease and parasites.

Causes. Malnutrition occurs most frequently when ducks are raised in buildings or grassless yards and fed nothing but grains or inadequate chicken rations. Unfortunately, mistakes are sometimes made in the manufacturing of feeds, and birds can be malnourished even though you are feeding them a supposedly adequate ration. Other causes include heavy parasite infestations and extremely hot weather, which reduces feed intake.

Treatment and Prevention. Provide an ample quantity of food that supplies a balanced diet. As temperatures go up feed consumption goes down, so in hot weather the nutritional density of feed must be higher for adequate nutrition. If you are trying to save money by using the cheapest feed available, keep in mind that per unit of nutrition, higher-priced feeds are often the best deal.

Omphalitis

When ducklings hatch, their navels are sometimes infected by microorganisms that cause omphalitis. Trouble can be signaled by inflamed navels or abdomens that are abnormally distended and mushy feeling. If not observed closely, ducklings with omphalitis can appear fairly normal until a short time before expiring. Infected birds in advanced stages usually huddle close to the heat source and move about reluctantly. Mortality, which may be light or heavy, invariably takes place between the second and sixth days after hatching.

Cause. It is inevitable that bacteria are present in nests and incubators, but poor sanitation increases the chance of omphalitis. Excessive humidity compounds the problem by slowing down the normal healing process of the navel and providing an ideal habitat for the bacteria.

Treatment and Prevention. Prevention is the only effective therapy. Providing an adequate number of nests that are generously furnished with clean nesting material is where prevention begins (granted, some ducks refuse to use nests!). Soiled hatching eggs should be washed with clean, lukewarm water to which an appropriate sanitizer, such as Germex, has been added. Finally, the incubator and hatcher must be kept clean and disinfected after each hatch. (Whenever possible, incubate eggs in one machine, then hatch them in a separate hatcher, thoroughly disinfecting it after each hatch.) Duck eggs should be candled on the 6th, 12th, 19th, and 24th days of incubation to avoid "blowouts" that spew millions of virulent bacteria into the incubator or hatcher.

Oviduct, Eversion of the *(Common)*

While attempting to lay eggs, females will occasionally expel a portion of their oviduct. A duck with this problem is easily identified by the expelled portions of the oviduct that protrude from her vent.

Causes. Possible causes include obesity, premature egg production, oversized eggs, excessive mating, and prolonged egg production. My experience indicates that the first three causes are by far the most significant.

Treatment. An ailing duck can be saved only if she is discovered relatively soon after the oviduct is dislodged and if prompt action is then taken. Carefully catch the bird, doing what you can to keep the oviduct clean. Gently wash the protruding oviduct with clean, lukewarm water to remove dirt and feces. Mineral oil can be applied to reduce drying of the tissues. Normally, if the oviduct is pushed back into place, it will come back out unless a few purse-string sutures are placed in the vent (this continuous stitch allows eggs and feces to pass without tearing out the sutures and is best installed by a veterinarian.)

During the recovery period the patient should be kept in a warm, dry, clean pen away from all males (a female companion may help keep her calm) and fed a nonlayer feed to discourage laying. Once the abdominal muscles have had time to strengthen, sutures can be removed (usually in 8 to 12 days). Full recovery is aided by isolation from drakes and discouraging laying for at least 3 months.

Prevention. Do not push young ducks into laying before they are eighteen to twenty weeks old, and make sure females are in good flesh, but not overly fat, at the beginning of and throughout the laying season. If excessive mating is a problem, drakes with overabundant libidos can be penned separately and put with the females for several hours twice a week.

Parasites, External

One of the characteristics of ducks that I have appreciated over the years is their general resistance to external parasites. It is common for healthy ducks that live outside to have incidental numbers of lice. However, during the half-century I have been raising ducks, there have only been a couple of times when the parasite load has been sufficiently high that treatment was warranted. If overdone or used recklessly, insecticides can cause more harm than good. Also, insecticides that come in contact with the vent can cause temporary sterility.

The first indication that a bird has external parasites can be repeated scratching of the head (especially in the area of the ears) and neck with the feet. If you look closely under a good light, you can usually see body, head, and neck lice with the naked eye by parting the feathers of the head, neck, and around the vent and oil gland. By holding an open wing up to a light source, such as the sun, you can see wing lice — if present — which are visible as dark lines ⅛ to ³⁄₁₆ inch (0.3 to 0.5 cm) long in the webbing of the secondary and primary flight feathers. Mites are often overlooked because of their tiny size, and they may be on the victim at nighttime only. In severe infestations mites sometimes are visible swarming over the surface of feathers or on the skin when feathers are parted and exposed to sunlight. After you have handled an infested bird, the miniature ticklike mites may also be seen and felt on your hands and arms.

Lice-infested ducks may have retarded growth, lose weight, lay poorly, and be agitated from the irritation of the lice. Heavy infestations of mites cause slow growth, anemia, weight loss, deserted nests, and even death.

Causes. When waterfowl are raised in a relatively clean environment and have access to bathing water, lice and mites do not often multiply enough to be troublesome (in some localities external parasites are much more prevalent than in others). However, ducklings that are hatched or brooded naturally or adult ducks that do not have access to swimming water can harbor harmful infestations of external parasites.

In general lice are small, flat, yellowish tan insects that normally live their entire life on the host bird. The most common species found on ducks include the wing louse, head-and-neck louse, and body louse. Mites are bloodsuckers for the most part. Some species, such as the feather mite and de-pluming mite, stay on the birds most of the time, while the red mite normally is on the birds only while feeding.

Treatment. Various treatments that have been suggested over the years include diatomaceous earth, olive oil, 50 percent organic apple cider and water solution, pulverized dried tobacco leaves, and wood ashes. There are various commercial insecticidal preparations available from feed stores or poultry supply distributors, including enzyme-containing lice & mite sprays, malathion, permethrin, and Sevin. To be effective these products need to be worked into the feathers of the head, neck, wings, upper tail, back, and vent. When dusting or spraying with an insecticide, carefully follow the manufacturer's directions. Extreme caution should be taken to avoid inhaling the insecticide, getting it in either your eyes or those of the birds, and contaminating water or feed. In case of heavy mite infestations, buildings, nests, and roosting areas should be cleaned, disinfected with an approved disinfectant, and then sprayed or dusted with the product of your choice. Under the guidance of a veterinarian, ivermectin can also be used to control most external parasites.

Prevention. To keep external parasites in check, provide sanitary living conditions, supply bathing water when possible, and treat birds before lice or mites are numerous enough to be harmful. Turkey and chicken hens used as foster mothers should always be treated for lice and mites before their maternal chores begin.

Parasites, Internal

Healthy waterfowl have reasonably good natural immunity to most internal parasites. In 50 years we have not needed to treat our farm-raised ducks for internal parasites. However, before bringing new stock to our farm, we treat them for internal and external parasites during quarantine.

An infestation of worms causes retarded growth, lowered feed conversion, reduced egg production, and increased susceptibility to disease. (In the case of gapeworms, birds often attempt to clear their windpipes by vigorously shaking their heads and coughing.) In severe cases the above symptoms are accompanied by weight loss, weakness, diarrhea, and, if not treated promptly, eventually death. Upon postmortem examination, most species of worms can be found in their appropriate habitat, if they are present in significant numbers.

Causes. Worms are primarily a problem when ducks have access to stagnant water, crowded ponds, or small streams or when forced to survive in a filthy environment. A variety of worms occur in poultry, including large roundworm, cecal worm, capillary worm, gizzard worm, gapeworm, and tapeworm.

Treatment. Treat ducks for internal parasites only if they actually have an infestation. The indiscriminate use of wormers over a period of time can reduce the natural immunity of the birds and develop vermicide-resistant parasites. The best way to know if your birds need treatment is to take a fecal sample to your veterinarian for evaluation.

Poultry worm medications to add to drinking water are readily available and easy to use. Some wormers work on only one species of parasite, while others are effective against several. Check with your feed or poultry-supply dealer for available brands, and follow instructions carefully. For a persistent problem a carefully planned worming schedule as outlined by the manufacturer will be needed to eradicate the parasites. Deworming ducks that are laying can negatively affect production. Some people have suggested diatomaceous earth or a pinch of chewing tobacco pushed down each bird's gullet as alternative remedies. Under the guidance of a veterinarian, ivermectin can also be used to control most internal parasites.

Prevention. Good sanitation practices, adequate diet, and the treatment and quarantine of all new arrivals to the farm are key to preventing heavy parasite infestations. Watering vessels should be placed on wire- or slat-covered platforms.

Pasteurella anatipestifer *Disease*

Also known as new duck disease, duck septicemia, and infectious serositis, *Pasteurella anatipestifer* disease is most common in two- to nine-week-old ducklings raised in dense concentration with poor sanitation and management. Symptoms in ducklings are similar to those in chickens with Newcastle disease: discharges from the nostrils and eyes, loss of appetite, green diarrhea, overall listless appearance, coughing, sneezing, head and neck tremors, and a stumbling gait. Ducklings that survive may develop swollen hock joints.

Cause. The bacterium *Pasteurella anatipestifer* is the cause; it has an incubation period of 1 to 5 days, with the first mortalities occurring from a few hours to a week after the first visible symptoms. Older birds are typically more resistant. Mortality can range from less than 10 percent to more than 75 percent. Good management practices reduce the mortality rate. *P. anatipestifer* is probably spread through the drinking water and inhalation of infected airborne particles.

Treatment. Birds showing severe symptoms normally do not respond to treatment. If caught early, a single injection of a combination of penicillin and streptomycin or sulfaquinoxaline in the feed or water normally reduces losses. (Streptomycin is toxic to some strains of birds, so it is a good idea to inject only one or two birds, then observe for 30 minutes. If they are sensitive, they will get droopy and stagger.) A good poultry vitamin mix added to the drinking water may help boost their immune systems.

Prevention. Good hygiene and management are the keys to prevention. New stock should be quarantined for 3 months, wild waterfowl kept away, and visitors kept out of pens. In areas where *P. anatipestifer* is common, some commercial producers use a bacterin vaccine.

Phallus Prostration (Common)

The penis, a 1½-inch- (3.75 cm) long (or longer) organ, protrudes from the bird's vent. Frequently, a drake with this disorder will be seen repeatedly shaking his tail from side to side as he attempts to retract the decommissioned organ.

Causes. Wild drakes normally pair off with a single duck and are sexually active for a relatively short period each spring. Under domestication, males often become sexually active at three and a half to five months of age, have multiple mates, and may breed the year round. Some drakes lose the ability to retract their penises, possibly caused by this unnaturally long and active mating season.

Treatment. Drakes with phallus prostration should be isolated immediately in a clean pen to prevent other ducks from damaging the penis by nipping it. If the bird is valuable and his problem is discovered before the penis has become infected or dried out, recovery is possible. The organ should be washed with clean, warm water, then disinfected and treated with a medicated ointment. You can try pushing the penis back into place, but — as often as not — it will pop back out in a short time. Apply the ointment daily until he is fully recovered, which may require several weeks or longer. Drakes should not be allowed to mate for at least 3 months.

Often, the end of the penis dries up and falls off. Because the semen exits from the base and travels down a groove that spirals around the penis, a drake can still be fertile even if a portion of the penis sloughs off (although fertility usually does not return for 6 to 12 months).

Prevention. The best safeguard is to keep backup drakes for breeding purposes.

Poisoning from Medication *(Common)*

Medication poisoning occurs either because the drug used is toxic to ducks or it was given in the wrong dosage. Symptoms vary depending on the age of the birds and the particular medication. Ducks may get sleepy, stagger, show signs of neck paralysis, lose their appetites, become weak, have stunted growth, or die suddenly.

Causes. In proportion to their size, ducklings consume more feed and water than chicks or turkey poults. Therefore, when given medications in their feed or water at the rate designated for chicks and poults, they are consuming overdoses. Also, some medications may be toxic to waterfowl but not land fowl. One prominent example is Ren-O-Sal, a medication commonly used for chickens that is deadly to ducklings. However, most modern medications, if given in the correct dosages for ducks, are safe.

Treatment. At the first sign of any of the above symptoms, discontinue the medication. If medicated feeds must be used, keep an eye out for symptoms of possible medication poisoning.

Prevention. Do not use Ren-O-Sal with ducks! When used in *duck-specific dosages*, medications that showed no negative effects in feed trials with ducks include amprolium, sulfaquinoxaline, Zoamix, bacitracin, chlortetracycline, dinitro-ortho-toluamide, ethopabate, furazolidone, novobiocin, and neomycin. Do not use medications except when a clear need is indicated.

Poisoning from Plants and Other Substances *(Common)*

Besides spoiled food and certain medications, other organic and inorganic substances can be poisonous to poultry if ingested in sufficient quantities. Diagnostic signs vary depending on the poison and the quantity ingested. In general, common symptoms at low or nonfatal levels include retarded growth, droopy appearance, and unsteadiness. At high levels birds may go into convulsions, fling their heads from side to side (apparently attempting to regurgitate the contents of their esophagus), or die suddenly. Due to the free gossypol content of untreated cottonseed meal, excessive feeding (more than 5 percent of the ration) of this protein supplement can result in suppressed growth, reduced egg production, and discolored egg yolks and albumin.

Causes. Some common materials that are known to be toxic to ducks include commercial fertilizers; salt; lead (from birds picking up lead pellets or nibbling on leaded paint); herbicides; pesticides; baits for rodents, slugs, and snails; leguminous plants and their raw seeds; cottonseed meal;

and leaves of tobacco and rhubarb. Other plants that have reportedly caused illness or death in ducks include foxglove, potato vines, potatoes that have turned green, castor bean plants and seeds, and eggplant leaves. This is only a partial list; there are many poisonous plants in various parts of the world.

Treatment. Treatment will depend on the particular circumstances and poison. Contact your veterinarian for treatment guidelines.

Prevention. Whenever using poisonous baits of any kind, locate them out of the reach of livestock. If commercial fertilizer is applied, do not spill or store it where ducks can get to it, and always make certain that the granules are dissolved for at least a week by rain or irrigation before birds are permitted on fertilized pasture. Buildings and equipment with peeling lead-based paint must be cleaned up or made inaccessible to ducks. Birds should not be allowed to eat sprayed or poisonous plants, salt, or icy slush resulting from snowy driveways or sidewalks being salted in winter. Cottonseed and alfalfa meals should not be included in a ration for ducks at more than 5 percent (for breeding ducks the use of cottonseed meal is questionable). Raw leguminous seeds, including raw soybean meal, should not be used. Even "safe" herbicide sprays should be used with great caution around ducks since they can cause decreased fertility in breeding birds.

Sinus Infection *(Common)*

See also chronic respiratory disease, page 276. Normally, the first noticeable symptom is a swollen sinus (between the bill and the eye), but nasal discharge, foamy eyes, and throat rattles may also be present.

Causes. Various microorganisms can cause sinus infections in ducks, but *Mycoplasma gallisepticum* is culpable in a majority of cases. The infection often lies dormant until a bird is stressed by factors such as shipping, moving to a new pen, sudden feed changes, drastic weather changes, or inadequate diet.

Treatment. At the first sign of a sinus infection, the affected bird should be isolated to minimize the possibility that it will infect other birds. There are many strains of mycoplasma and they vary in their sensitivity to different antibiotics. Tylosin (sold as Tylan) often is by far the most effective, followed by erythromycin (Gallimycin) and tetracyclines (Aureomycin and Terramycin).

Prevention. The best way to prevent sinus infections is to acquire mycoplasma-free stock. Unfortunately, birds that have had sinus infections can be carriers even after symptoms have long since disappeared.

Spraddled Legs (Common)

Ducks may have legs that slide out from under them as they attempt to walk. In acute cases the legs may protrude at right angles from the body. It is most often seen in newly hatched ducklings or in adults of the largest breeds and is usually easily remedied if quickly treated.

Causes. Almost all cases of spraddled legs are the direct result of birds being on slick surfaces where they have poor footing, including incubator trays, brooder floors, plastic pet carriers, and cardboard boxes. Ducks of all ages, but particularly ducklings, can be flipped onto their backs and have a difficult time righting themselves — especially if they happen to end up in a slight depression in the bedding or ground. If they spend too much time on their back flailing with their legs, spraddled legs are often the result.

Treatment. Ducks of any age that develop spraddled legs should be hobbled with an appropriate length of soft yarn that keeps the legs from separating any further than their natural distance. My preferred method is to place a cable tie (the size should be appropriate to the size of the bird) on each leg just above the back toe. The cable tie is tightened just enough so that it will not slip off the foot or up over the hock joint. The tail of the cable tie on each leg is cut flush with the lock of the cable tie. Each end of the hobble yarn is then tied to one of the cable ties. Place the patient in an area with very "grippy" flooring, such as wood excelsior pads or firmly packed hay. The length of time the hobble needs to remain in place ranges from an hour to a day or more, depending on how severe the muscle strain. When hobbles are used, it is essential that blood circulation is not restricted.

Spraddled legs (left) can usually be remedied if legs are hobbled (right) appropriately and in a timely matter.

Prevention. Never hatch ducklings on slippery incubator trays, or house or transport ducks of any age in an area or carrier with slick floors. In our hatching trays, we use woven rubber mats. An appropriate bedding should be used to provide good footing, and ducklings especially should be kept in an area without depressions so they cannot become stuck on their backs.

Staphylococcosis

In ducks the most common places for staph infections are the joints of the feet and legs. In the mild form birds show lameness and an unsteady gait, stand or sit by themselves, move about reluctantly, lose weight, and, if not treated promptly, eventually die. In severe cases the above symptoms can be accompanied by hot, swollen joints, diarrhea, and sudden death. Postmortem exams generally show congested and swollen livers, spleens, and kidneys.

Causes. The organism that causes staph infections, *Staphylococcus aureus*, can be found in most flocks of poultry, but it does not seem to be a serious threat unless ducks are in a run-down condition due to poor nutrition, parasites, injuries, or from being kept in grossly unsanitary quarters.

Treatment. Depending on the strain of staph, various antibiotics can be effective, including Gallimycin, Terramycin, novobiocin, and penicillin. (Prescription antibiotics can be obtained from your veterinarian.)

Prevention. When ducks are kept in reasonably sanitary conditions and adequately fed, staph is usually not a major problem. In brooding pens fresh bedding (we use coarse cedar sawdust) should be sprinkled on top before pens get greasy. During rainy weather the duck yard should not be allowed to deteriorate into filthy, mud-covered cesspools. Birds should not be allowed to walk on sharp objects that may puncture their feet, thus inviting this infection.

Sticky Eye *(Common)*

In this condition the feathers around the eyes are matted, and in severe cases the eyelids stick together.

Causes. Under domestication waterfowl are sometimes raised on diets deficient in such vitamins and minerals as vitamin A, pantothenic acid, and biotin. They also may be supplied water in shallow containers that do not permit them to rinse their eyes. The result is that ducks raised in confinement are sometimes susceptible to sticky eye. Normally, sticky eye is not a health problem if it is simply a matter of insufficient bathing water.

Treatment and Prevention. You can keep sticky eye to a minimum by supplying a healthful diet and adequate, clean bathing water. (Some ducks will keep their faces clean with access to a dishpan, whereas others need a larger pool.)

Streptococcosis

Under most circumstances streptococcosis is uncommon in ducks. The strep infection found in ducks is caused by the bacterium *Streptococcus gallinarum* and can be introduced into flocks by carrier birds. This disease is difficult to diagnose and is probably occasionally mistaken for fowl cholera because of its similar symptoms, including the sudden death of birds that appear to be in good health. The liver and lungs often show congestion and enlargement. Streptococcus must be isolated in the laboratory for positive identification.

Treatment and Prevention. Treatment is difficult because birds often die before infection is suspected. Antibiotics that are useful against gram-positive organisms (check with your veterinarian) have been suggested as a preventive measure and as a treatment for infected fowl when used at the highest recommended level. Acquiring healthy stock that has been raised in a clean environment and providing sanitary living conditions seem to be the best prevention.

Sudden Death

In this case a duck that a short while earlier appeared normal and healthy is found dead.

Causes. There's a host of causes, including heart attack, choking, hemorrhaging, blood clots, broken egg in oviduct, suffocation, exposure to toxic gases, consumption of toxic substances, being kicked or stepped on by a larger animal, attack by a predator (puncture wounds can be almost impossible to detect on heavily feathered portions of the body), heatstroke, seizure, broken neck, virulent disease organisms, and drowning. If ducklings are held so tightly that their bodies cannot expand and contract for breathing, they can be suffocated quickly.

Prevention. Some sudden deaths cannot be prevented. The rate of sudden deaths can be reduced by a proper environment, including providing plenty of shade and cool drinking water during hot weather; giving protection from large animals (that might step on them) and predators; avoiding situations that terrify ducks or cause them to panic; keeping the correct ratio of males to females (drakes may gang up on females and drown them while attempting to mate on water); and always using proper methods for catching and holding ducks. Good sanitation and management reduce the chances of infectious diseases that cause sudden death.

West Nile Virus

Over the last decade the West Nile virus has been spreading across North America and can infect birds, humans, and various other mammals. In ducks the symptoms most commonly reported are neurological in nature. The first time someone contacted me with a case of West Nile, the symptoms the person described sounded vaguely like a mild form of botulism. However, the symptoms did not follow the distinctive pattern of botulism, so I recommended that the person have the affected birds tested for both botulism and West Nile. The lab test came back positive for West Nile.

In the mildest confirmed cases reported to me, infected ducks tend to be lethargic and lag behind the flock, may show slight head tremors, and may act mildly disoriented. In more severe cases symptoms include difficulty in holding up the head, severe staggering or total inability to walk, and eventually death. Unlike in botulism, where infected ducks tend to either die within a few hours or show noticeable improvement over the course of a day, the ducks infected with West Nile tend to have lingering symptoms and do not respond to treatments that are effective for botulism.

Causes. The West Nile virus is primarily spread by mosquitoes, but it can also be transmitted by other biting insects. It is not believed that the virus can be spread directly from bird to bird.

Treatment and Prevention. At this point there is no vaccination available for birds and no known treatment. Patients should be made as comfortable as possible and given easy access to feed and drinking water, to which a water-soluble vitamin and electrolyte supplement can be added. Prevention includes minimizing standing water where mosquitoes can breed. Some duck breeders who have had West Nile affect their flocks have found that raising their ducklings in tightly screened, mosquito-proof enclosures seems to reduce the incidence of West Nile infections.

Wing Disorders (Common)

There are three main wing disorders: twisted wing (the wing tip sticks out from the body when the wing is folded), split wing (there is a gap between feathers in the wing), and lazy wing (the tip of the wing hangs down along the side of the body). These are primarily an aesthetic concern and seldom affect productivity.

Causes. The most common cause is a combination of improper diet and insufficient exercise. Improper diet can include excessive or inadequate protein, vitamins, and minerals. In general diets containing more than 16 percent protein after ducklings are two weeks old greatly increase wing disorders. Ducklings raised outside with access to swimming water and an appropriate diet seldom have wing disorders. Other causes include excessively high brooding temperatures, feather eating, broken wing feathers, narrow feathers (common in Runners), injuries, and genetic defects.

Treatment. When treated before the bones of the wing harden into position, it is often possible to repair twisted wings of ducklings by manually folding the feathered limb in the correct position and taping it shut with ½-inch (1.25 cm) masking tape. Because the wing will become stiff from nonuse, the tape should be removed after 10 to 14 days and, if need be, retaped after half a day of exercise. For mature birds that cannot be rehabilitated, clipping the flight feathers improves their appearance.

Prevention. Wing disorders can be greatly reduced by feeding balanced diets with no more than 18 percent protein from zero to two weeks and 15 to 16 percent protein up to ten weeks. Ducklings should be allowed outside to get exercise and swim as soon as it is safe.

Twisted (also called slipped) wings can sometimes be fixed by temporarily taping them in position.

PRECAUTIONS WHEN USING DRUGS AND PESTICIDES

When used according to the instructions and along with good management practices, the occasional use of drugs and pesticides can be useful in maintaining the health and productivity of poultry flocks. However, if overused or misused, these aids pose health hazards to both animals and humans. Always follow directions carefully and use only the dosages recommended for the specific problem. Don't fall into the trap of thinking that if one dose is good, then a double dose must be better. To keep from poisoning yourself and your family or customers with potentially dangerous drugs, follow recommended withdrawal periods when treating meat- or egg-producing birds. And last but not least, store all drugs and pesticides in their original containers in a dry, clean location out of the reach of children and animals.

Pullorum-Typhoid Blood Testing

Pullorum and fowl typhoid are two types of highly contagious *Salmonella* infections that can be passed from breeding stock to offspring through hatching eggs or by infected birds coming in contact with healthy birds. Since these diseases are easily transported from one locality to another, some states (and most countries) require that all adult poultry crossing their borders, either for breeding stock or for exhibiting in a show, must be blood tested for pullorum-typhoid and certified clean. For transporting or shipping day-old poultry, the parent stock must be certified clean.

For the small-flock owner, blood testing is not usually required (although it's not a bad idea and is relatively inexpensive in states that cooperate with the National Poultry Improvement Plan) unless birds are to be exhibited or sold in states requiring a health permit. There is some risk if you acquire stock from breeders who do not blood test.

For waterfowl and chickens the pullorum-typhoid test is normally performed by extracting a small amount of blood from under the bird's wing and mixing it with a drop of antigen on a light table. To be valid, blood tests must be performed by a licensed technician. For more information contact the veterinary office of your state or province Department of Agriculture.

Postmortem Examination

Most of us who raise poultry have experienced the disappointment of finding an expired bird that was in good flesh and showed no outward signs of disease or attack from a predator. When this occurs, you can (1) dispose of the carcass and never know what caused the bird's demise, (2) take the fowl to a diagnostic laboratory or veterinarian for diagnosis, or (3) perform a postmortem yourself and see if the problem can be located.

The first of the three choices is probably the most commonly chosen but definitely the least desirable. The second is most desirable but least common and sometimes impractical, except when a serious or contagious disease is suspected. While laypersons cannot approach the proficiency of trained diagnostic specialists, the third alternative can be useful in identifying some problems to avert further mortality or lowered production.

Equipment Needed

The average household contains the few tools needed to perform a basic postmortem examination. A small, sharp knife (which afterwards should not be used on human food) is needed for opening the bird. A pair of scissors is useful for incising the trachea (windpipe), esophagus, intestines, and ceca. A magnifying glass is helpful in searching for internal parasites, such as hard-to-see capillary and cecal worms. The work area should be covered with plastic and located in a well-lit area.

Caution: After a postmortem all tools must be sterilized and the work area and your hands disinfected to guard against spreading infections and disease to humans and livestock.

Procedure for a Basic Postmortem

When examining waterfowl I have adapted the following procedure, although the sequence of the examination may vary. If an esophagus or gizzard impaction is suspected, these organs are inspected first; if worms are a prime candidate, then the intestines are examined first, and so on.

1. Cut and spread open the bird from one corner of the mouth to the vent.
2. Scan the body cavity and organs for hemorrhages and tumors.
3. In mature females examine the body cavity and oviduct for abnormalities such as internally laid eggs, blocked oviduct, or obesity.
4. Examine the liver for discoloration, tiny light spots, light streaks, dark hemorrhagic areas, hardness, or yellow coating of waxy substance.

5. Cut open the gizzard, and check for sharp foreign objects, a hard mass of string or other material, eroded inner lining, or serious damage. Peel off the horny inner lining and look for uncharacteristic bumps, which may indicate gizzard worms.

6. Open the esophagus and look for blockage or injury.

7. Examine the main organs (liver, heart, spleen, lungs, kidneys, and ovaries) for obvious deformities, discoloration, or infections.

8. Slit open the small and large intestines and ceca, checking for worms; blood; inflamed linings; hemorrhages; and yellow, cheesy nodules.

9. Split the trachea lengthwise and inspect for blockage, gapeworms, blood, excess mucus, and cheesy material.

10. When finished with the examination, clean up and dispose of the carcass and all debris, disinfect the tools, and thoroughly wash your hands and arms with soap and warm water.

11. To maximize the effectiveness of the postmortem examination, you can write up a short report of your findings for future reference.

18

Showtime

NORTH AMERICAN POULTRY BREEDERS have been displaying their fowl at public exhibits for over 150 years. By digging into the annals of poultry history, we find that the first exclusive poultry show was held on November 14, 1849, at the Public Gardens in Boston, Massachusetts. The Boston Poultry Exposition continues as an annual event.

Poultry shows serve a number of useful functions. In addition to providing a meeting place for old and new friends who share a common interest, they allow the public to see and enjoy the breeders' skills and provide the opportunity to compare one's stock and ideas with those of other breeders. Shows can also be excellent places to advertise and sell your birds, and many people enjoy the competition of exhibiting their ducks.

Kinds of Shows

Shows come in many sizes and shapes. There are county, state, provincial, club, and national shows, ranging in size from a few dozen birds to several thousand — the largest having topped 10,000 entries. At county and state fairs, the open-class division can be entered by anyone, whereas the 4-H and Future Farmers of America (FFA) divisions are limited to members. Club shows are usually divided into youth division (for anyone eighteen years and younger) and open class (for exhibitors of all ages).

Although shows were originally held primarily in the fall, today they may be held in any season of the year. For information on show dates and locations, call local fairgrounds, check with local poultry clubs, and subscribe to the *Poultry Press* and *Feather Fancier*, monthly newspapers that carry advertisements for most club shows (see appendix H, page 341).

Open Class

In the open-class division, most ducks are judged in their show cages as the judge walks along the aisle. Unlike the way chickens are judged, duck judges often do not take the birds out of their cages to minimize the possibility of injuries. At some shows Runners are judged in a show ring and the owner or show assistants carry the birds from their cages to the show ring at the appointed time.

4-H, FFA, and Youth Showmanship Class

In most 4-H and FFA shows, birds are carried by their owners or a helper to a show "bench" where they are presented to the judge for evaluation. In conformation classes the birds are judged by the prevailing breed standard. In showmanship classes, the primary emphasis is on the knowledge and appearance of the owner and the behavior and cleanliness of the bird on the show bench. For showmanship most fairs do not discriminate against crossbreds or mediocre purebreds. County Extension agents normally have material available on preparing birds and exhibitors for showmanship classes.

Who Can Show?

The majority of poultry shows are open to everyone who wants to enter a bird. Belonging to a local or national organization is not usually a prerequisite for exhibiting your ducks, although some county and state fairs limit entrants to residents. To help cover expenses, there is a reasonable entry fee. In some states any fowl that is shown from out of state must be blood tested for pullorum-typhoid; a few states require the test for in-state birds as well. Premium lists spell out such requirements, and the show secretary can provide information on how to obtain the blood test. (Pullorum-typhoid testing, if required, is usually provided at the show.)

What Can Be Shown?

Any healthy duck of a standard breed can be exhibited at shows sanctioned by the American Poultry Association (APA). Most shows will also accept non-standard breeds, although they cannot compete for awards against standard breeds. At some county and state fairs, there are also classes for crossbreds.

These ducks are part of the Champion Row at the 1999 Northwest Winter Classic.

How Are Ducks Entered?

To exhibit ducks in a show, you need a copy of the premium list and an entry blank. For county, provincial, and state fair premium lists and entry blanks, contact your local agriculture Extension specialist. For other shows check with established poultry breeders in your area, and look over advertisements for shows in the *Poultry Press* or *Feather Fancier*. Normally, entry forms must be filled out and returned to the show secretary anywhere from 1 to 4 weeks prior to the show date, so you'll need to plan ahead. To avoid being disqualified be sure to read and follow all instructions as outlined in the premium list.

Shows normally have separate classes for old drakes, old ducks, young drakes, and young ducks. Be careful to enter birds of the correct sex and age in the appropriate class. Some shows require birds to be tagged with numbered leg bands, and these numbers are recorded on the entry blanks.

Selecting Show Ducks

Selecting show birds is a combination of art and science. To become proficient at selecting show ducks, it is necessary to learn the desired characteristics of your chosen breed. A good starting point is to study the section on breeds in this book and the *American Standard of Perfection*. Then make a list of the most important three or four traits of the breed. For example, a list of the main characteristics for Call ducks would include:

1. Small size
2. Short bill
3. Large, round head
4. Wide, plump body

Novices often place too much emphasis on minor details that normally have little or no bearing on a bird's value as a show specimen. Examples of minor details that are often overemphasized include toes that are not perfectly straight (it's normal for the outside toe in particular on each foot to be somewhat curved) and minor variations in foot, bill, eye, or plumage color.

Keep in mind that type is more important than color in show birds. In the *American Standard of Perfection*, on a 100-point scale, color counts for 29 points in most white ducks and for 36 points in most color varieties. A bird with excellent type and size but of average color will typically beat a bird with average type and size but excellent color.

At many shows a judge will appraise between 200 and 500 birds in a single day, meaning he or she can only spend a short time evaluating each bird. Winning ducks are strong in their breeds' distinctive characteristics, and as celebrated judge Henry Miller often said, their "quality meets the eye" as the judge walks down the aisle.

An excellent way to hone your evaluation skills is to prejudge your ducks and then compare against the judge's placement at the show. Most judges, once they are finished judging for the day, are willing to explain why they placed a class of birds as they did.

Preparing Ducks for Exhibition

To show to their best advantage, ducks must be clean, in good feather, carrying the correct amount of weight, and accustomed to being penned in a restricted enclosure. Birds that are dirty, in poor feather condition, or are over- or underweight will fare poorly under most judges, even if they are excellent specimens otherwise. Also, while appraising ducks, judges have little patience with birds that thrash wildly about or crouch in the corner of their cages.

Cage Training

For ducks that have always had the freedom to run around in a spacious pasture or yard, being locked up in a display cage can be a disconcerting experience. The first phase in training ducks for exhibition is to work calmly when near the birds and talk to them with a reassuring voice at feeding time. The calmest ducks are those that their owners have worked with from an early age. At a minimum, several weeks prior to the show date, it's a good practice to coop birds that have never been exhibited in wire cages for 2 or 3 days with food and drinking water. The cages should be at least 24 inches (60 cm) wide × 24 inches deep × 27 inches (69 cm) tall for heavy breeds (cage size can be adjusted down for smaller breeds) and located where you will walk by them occasionally throughout the day so the ducks become accustomed to people trafficking by on foot.

Conditioning

In shows ducks are judged on a 100-point scale. The APA *Standard of Perfection* allots 10 of these points for condition and vigor. In good competition the winners are clean and in excellent feather condition. Fortunately, ducks are neat birds and will clean themselves if given half a chance. Seldom do they need to be hand-washed prior to a show, except possibly to tidy the bill, feet, or a few soiled feathers with warm water and a soft brush (an old toothbrush is great for cleaning bills and feet) or sponge. Three of the worst enemies of show ducks are mud, overcrowding, and excessive exposure to the sunlight, which can cause the plumage to fade, dry out, and lose its sheen.

To encourage good condition, ducks should be kept in a clean environment, provided ample shade, and fed a diet that encourages lustrous feathering for 6 to 12 weeks prior to a show. (For more details on conditioning, see specific breed recommendations.)

Fitting Show Birds

Grooming birds for the show pen is an enjoyable and relaxing activity for many waterfowl exhibitors. Six to 8 weeks prior to an exhibition, show prospects can be examined for broken, stained, or faded feathers. If faulty feathers are plucked at this time, they should be regrown in time for the show. A day before the show, any small, off-colored body feathers can be removed, the toenails clipped and filed, the feet and bill washed with warm water, and a clean pan of bathing water provided in the pen for last-minute bathing. (As they are catching their ducks for a show, some people rub baby oil or some other substance onto the legs and feet, but great care must be taken to not get any on the feathers.)

Faking

Because poultry shows are partly beauty contests, exhibitors want their birds to look their best. However, artificially coloring any part of the bird and the cutting or removal of main wing and tail feathers are considered faking and can result in the disqualification of a bird.

BIRD OWNERSHIP

Ducks must be owned by the person or farm under whose name the entry is made. 4-H, FFA, and some specialty club shows sometimes require a minimum length of ownership time prior to showing. However, most open-class shows have no minimum-length-of-ownership requirements.

On the other hand, the removal of off-colored body feathers (such as white feathers in a black duck) or the feeding of special feeds to enhance the color of feathers (for example, feeding yellow-pigmented ingredients to Pekins to increase the creamy yellow of their plumage) or bill are part of showing your ducks to their best advantage and are not considered faking.

Transporting Ducks to Shows

Ducks arrive at shows in many different kinds of containers, from cardboard boxes to fancy custom-designed carriers. The main concern in getting your birds to the show is that they have sufficient ventilation, clean bedding, and adequate space. For Runners, I use crates that are a minimum of 28 inches (70 cm) tall to prevent them from bumping their heads, which can cause temporary neck kinking. An advantage of cardboard boxes is that they can be recycled when they are soiled; however, be sure to cut sufficient holes to provide good ventilation. Wooden crates are sturdier and can be decorated with the farm name or colors.

Care at the Show

Some county and state fairs provide attendants who feed and water the birds throughout the duration of an exhibition. However, club shows do not normally provide this service, and you are responsible for feeding and watering your entries. If this matter is not clarified in the premium list, you should check ahead of time so you will not be caught without feed and containers if this service is not provided.

Showroom Etiquette

Most poultry shows are operated by volunteers who put in long hours and often significant amounts of their own money to make the shows happen. Whenever possible, thank the host club members and show workers for their efforts and ask if there is any way you can help. If everyone pitches in, shows become more enjoyable for all.

Unless judges initiate conversation, exhibitors should not communicate with them or in any way interfere during judging. Many shows allow spectators to observe the judging from a reasonable distance; however, observers should be quiet and not make loud comments about the quality or ownership of the birds.

HOMEMADE SHOW AND UTILITY CAGES

Handy wire cages — for show training, hospital isolation, quarantine, breaking up broody ducks, and other uses — can be made at home. Tools needed are wire-cutting pliers, cage clip applicator pliers, cage clips, 3-foot- (1 m) tall welded wire fencing with 1 × 2-inch (2.5 × 5 cm) openings, and a flat metal file.

To make a single cage that is 3 feet long, 2¼ feet (0.75 m) deep, and 3 feet tall (this size will accommodate even the tallest Runner and is easy to catch and remove birds from), purchase 15 feet (4.5 m) of 3-foot-high welded wire and a 2-inch hasp (clip).

Cut an 11-foot (3.5 m) length of wire and bend it at right angles after 36 inches (1 m), then another 30 inches (0.75 m), and then another 36 inches to make the corners of the cage. The fourth corner is made by clipping together the two loose ends with half a dozen cage clips. For the top, cut a 30-inch length and clip it all the way around with cage clips.

On one of the 3-foot- (1 m) long sides, measure up 6 inches (15 cm) from the bottom and cut a 16-inch- (40 cm) tall by 14-inch- (35 cm) wide opening in the center of the cage. (With the metal file smooth any jagged wire ends so your arms will not be scratched as you take birds in and out of the cage.)

From the remaining original piece of wire, cut a door that is 20 inches (50 cm) tall and 16 inches (40 cm) wide. Position the door over the opening so that it overlaps 1 inch (2.5 cm) on each side and 2 inches (5 cm) on top and bottom, and clip along one side. Use the 2-inch hasp to secure the door shut.

This cage can be set on a piece of plywood and bedded with wood shavings or straw, or welded wire with ½ × 1-inch (1.25 × 2.5 cm) openings can be clipped to the bottom for a wire floor.

Once judging is completed most judges are willing to discuss why they placed the birds the way they did. This is not a time, however, to argue or attempt to change the placing of a bird. Remember, when you enter a bird in a show, you are agreeing to the appraisal of the hired judge. People who are good sports learn the most and receive the greatest enjoyment from shows and will congratulate the winners regardless of their personal feelings.

Interpreting Judging

One of the benefits of showing is having a disinterested person evaluate your birds. It is best not to take too seriously either the pros or cons of a single judge; even highly respected judges can disagree on the merits of a given bird. If breeders base decisions about which birds to breed from strictly on how they place at a show, their breeding programs may be hampered.

I have watched many people dispose of valuable birds based on the evaluation of one judge. Once at a national waterfowl meet, I was standing with several junior exhibitors during the judging of a class of ducks that they were exhibiting. The judge made a number of derogatory comments about their birds, which prompted the young exhibitors to say that they were going to dispose of them. I suggested that their birds were of good quality and that they should show them at an upcoming show where a nationally known judge, more familiar with this breed, would be judging.

At the second show I again stood with the young exhibitors as their birds were being evaluated. This time the judge specifically commented on how these birds were better than most he had seen of this relatively rare variety — two well-known judges and two very different evaluations.

Young people can raise world-class ducks. Rebecca Simon shows off her champion Crested duck.

SHOW TERMS AND ABBREVIATIONS

OF (old female): Ducks over one year old

OM (old male): Drakes over one year old

YF (young female): Ducks under one year old

YM (young male): Drakes under one year old

BB or **BOB**: Best of Breed

BOSB: Best Opposite Sex of Breed (same breed as BB)

BV or **BOV**: Best of Variety

BOSV: Best Opposite Sex of Variety

ASV: All Standard Varieties

Disq.: Disqualified

General Show Hints

Even out of the best breeding stock, only a portion of the offspring will be show specimens. When raising ducklings for show, plan to raise a minimum of two or three for each show bird desired. Most experienced breeders find that they must hatch anywhere from 15 to 100 ducklings to produce an elite show bird that is capable of taking top honors in the stiffest competition.

Although raising large numbers of ducklings generally increases your chances of producing an elite show bird, you should never raise more than you are capable of caring for properly. High-quality show birds are the result of good breeding, correct diet, and a healthful environment.

APPENDIX A

Mixing Duck Rations

For breeders who would like to mix their own duck rations, here are some examples of balanced feeds for ducks at various stages of development.

HOME-MIXED STARTING RATIONS (0–2 WEEKS)

Ingredient	Ration 1 Small Quantity	Ration 2 Large Quantity
Yellow cornmeal	11 cups (2.6 L)	62 lbs (28 kg)
Soybean meal (44% protein)	3½ cups (830 ml)	17 lbs (7.7 kg)
Wheat bran	2 cups (475 ml)	2 lbs (0.91 kg)
Meat and bone meal (50% protein)	½ cup (120 ml)	4 lbs (1.8 kg)
Fish meal (60% protein)*	½ cup (120 ml)	2 lbs (0.91 kg)
Alfalfa meal (17.5% protein)	½ cup (120 ml)	2 lbs (0.91 kg)
Dried skim milk or Calf-Manna	½ cup (120 ml)	3 lbs (1.4 kg)
Dried brewer's yeast	1½ cups (350 ml)	7¼ lbs (3.3 kg)
Dicalcium phosphate (18.5% phosphate)	1 Tbsp (15 ml)	½ lb (225 g)
Iodized salt	1 tsp (5 ml)	¼ lb (110 g)
Cod liver oil**		
Totals	20 cups (4.4 L)	100 lbs (45.4 kg)
Chopped succulent greens	Free choice	Free choice
Sand or chick-sized granite grit	Free choice	Free choice
Chick-size oyster shells or crushed dried eggshells	Free choice	Free choice

*If fish meal isn't available, use 4½ cups (1 L) soybean meal and 10½ cups (3 L) cornmeal in formula #1 or 20 lbs (9 kg) soybean meal and 61 lbs (28 kg) cornmeal in formula #2.
**If birds do not receive direct sunlight, which enables them to synthesize vitamin D, sufficient cod liver oil must be added to these rations to provide 500 international chick units (ICU) of vitamin D_3 per pound of feed.

HOME-MIXED GROWING RATIONS (2–12 WEEKS)

Ingredient	Ration 3 Corn Base in Lbs (kg)	Ration 4 Wheat Base in Lbs (kg)
Cracked yellow corn*	75 (34)	—
Whole soft wheat	—	79 (36)
Milo (grain sorghum)	—	—
Soybean meal (50% protein)	9 (4.1)	5 (2.3)
Meat and bone meal (50% protein)	4 (1.8)	4 (1.8)
Alfalfa meal (17.5% protein)	3 (1.4)	3 (1.4)
Dried skim milk	3 (1.4)	3 (1.4)
Dried brewer's yeast	5 (2.3)	5 (2.3)
Dicalcium phosphate (18.5% phosphate)	0.10 (0.05)	0.10 (0.05)
Limestone flour or oyster shells	0.65 (0.3)	0.65 (0.3)
Iodized salt	0.25 (0.11)	0.25 (0.11)
Cod liver oil**		
Total (lbs)	100	100
Sand or chick-size granite grit	Free choice	Free choice
Chopped succulent greens (eliminate when pasture is available)	Free choice	Free choice

*Whole corn can be used after the birds are four to six weeks of age.
**See Home-Mixed Starting Rations (0–2 Weeks).

HOME-MIXED RATIONS FOR ADULT DUCKS

Ingredient*	Ration 5 Holding in Lbs (kg)	Ration 6 Holding in Lbs (kg)	Ration 7 Layer in Lbs (kg)	Ration 8 Layer in Lbs (kg)
Whole milo or yellow corn	82 (37)	—	60 (27)	24 (11)
Whole soft wheat	—	86 (139)	9 (4.1)	48 (21.8)
Soybean meal (50% protein)	8 (3.6)	4 (1.8)	7 (3.2)	4.5 (2)
Meat and bone meal (50% protein)	—	—	4 (1.8)	4 (1.8)
Alfalfa meal (17.5% protein)	4 (1.8)	4 (1.8)	4 (1.8)	4 (1.8)
Dried skim milk	—	—	2 (0.9)	2 (0.9)
Dried brewer's yeast	5 (2.3)	5 (2.3)	7 (3.2)	7 (3.2)
Oyster shell	0.25 (0.11)	0.50 (0.25)	6.4 (2.9)	5.9 (2.7)
Dicalcium phosphate (18.5% protein)	0.50 (0.25)	0.25 (0.11)	0.30 (0.15)	0.35 (0.15)
Iodized salt	0.25 (0.11)	0.25 (0.11)	0.30 (0.15)	0.25 (0.11)
Cod liver oil**				
Totals (lbs)	100 (45)	100 (45)	100 (45)	100 (45)

*The addition of 3 to 5 pounds (1.3 to 2.2 kg) of livestock-grade molasses to each 100 pounds (45.5 kg) of mixed feed reduces waste.

**If birds do not receive direct sunlight, which enables them to synthesize vitamin D, sufficient cod liver oil must be added to these rations to provide 400 international chick units (ICU) of vitamin D_3 per pound of feed.

Note: These rations are best suited for situations when birds have access to pasture.

COMPLETE RATIONS FOR DUCKLINGS (PELLETED)

Ingredient	Ration 9 Corn Base Starter in Lbs (kg)/Ton	Ration 10 Wheat Base Starter in Lbs (kg)/Ton	Ration 11 Corn* Base Grower in Lbs (kg)/Ton	Ration 12 Wheat* Base Grower in Lbs (kg)/Ton
Ground yellow corn	1385 (628)	—	1570 (712)	—
Ground soft wheat	—	1425 (646)	—	1607 (729)
Ground milo (grain sorghum)	—	—	—	—
Soybean meal solv. (50% protein)	430 (195)	360 (163)	295 (134)	270 (122)
Fish meal (60% protein)**	40 (18)	40 (18)	—	—
Meat and bone meal (50% protein)	80 (36)	80 (36)	80 (36)	40 (18)
DL-methionine (98%)	2 (0.9)	2.75 (1.25)	1.5 (0.7)	2 (0.9)
Stabilized animal fat	25 (11.3)	40 (18.1)	—	20 (9.1)
Soybean oil	—	14 (6.4)	—	10 (4.5)
Dicalcium phosphate (18.5% phosphate)	6.25 (2.8)	5.25 (2.4)	8 (3.6)	17 (7.7)
Limestone flour (38% calcium)	6 (2.7)	7 (3.2)	20 (9.1)	9 (4.1)
Iodized salt	5.75 (2.6)	6 (2.7)	5.5 (2.5)	5 (2.3)
Vitamin:mineral premix	20 (9.1)	20 (9.1)	20 (9.1)	20 (9.1)
Totals (lbs)	2000 (907)	2000	2000	2000
VITAMIN:MINERAL PREMIX				
Vit. A (millions of IU/ton)	8.0	9.5	5.0	6.5
Vit. D_3 (millions of ICU/ton)	1.0	1.0	0.8	0.8
Vit. E (thousands of IU/ton)	5.0	15.0	2.0	12.0
Vit. K (g/ton)	2.0	2.0	2.0	2.0
Riboflavin (g/ton)	6.0	6.0	4.0	4.0
Vit. B_{12} (mg/ton)	8.0	8.0	4.0	4.0
Niacin (g/ton)	50.0	50.0	40.0	40.0

COMPLETE RATIONS FOR DUCKLINGS
(PELLETED) (*CONTINUED*)

Ingredient	Ration 9 Corn Base Starter	Ration 10 Wheat Base Starter	Ration 11 Corn* Base Grower	Ration 12 Wheat* Base Grower
d-calcium pantothenate (g/ton)	6.0	3.0	4.0	2.0
Choline chloride (g/ton)	300.0	—	200.0	—
Folic acid (g/ton)	0.5	0.5	0.5	0.5
Manganese sulfate (oz/ton)	9.6	9.6	9.6	9.6
Zinc oxide (80% zinc, oz/ton)	3.2	3.0	3.2	3.0
Ground grain to make 20 lbs	+	+	+	+
Totals (lbs)	20.0	20.0	20.0	20.0
CALCULATED ANALYSIS				
Crude protein (%)	20.2	19.6	16.4	16.1
Lysine (%)	1.05	1.00	0.76	0.73
Methionine (%)	0.46	0.44	0.37	0.33
Metabolizable energy (kcal/lb)	1424	1366	1426	1362
Calorie:protein ratio	70	70	87	85
Crude fat (%)	4.7	4.7	3.5	3.3
Crude fiber (%)	2.3	2.4	2.3	2.4
Calcium (%)	0.76	0.78	0.93	0.63
Available phosphorus (%)	0.41	0.41	0.36	0.36
Vit. A (IU/lb)	4912	4750	3535	3250
Vit. D$_3$ (ICU/lb)	500	500	400	400
Riboflavin (mg/lb)	3.73	3.76	2.64	2.66
Total niacin (mg/lb)	41	51	31	43

*To convert into breeder-developer rations, substitute the following quantities of vitamins for those listed under the vitamin:mineral premix: 1 g of vit. K; 3 g riboflavin; 30 g niacin; and 100 g choline chloride in the corn and milo base rations.

**If fish meal is not available, use an additional 50 pounds (23 kg) soybean meal solvent (50% protein) and subtract 10 pounds (4.5 kg) of grain.

COMPLETE RATIONS FOR ADULT DUCKS (PELLETED)

Ingredient	Ration 13 Corn Base Holding in Lbs (kg)/Ton	Ration 14 Wheat Base Holding in Lbs (kg)/Ton	Ration 15 Corn Base Breeder in Lbs (kg)/Ton	Ration 16 Wheat Base Breeder in Lbs (kg)/Ton
Ground yellow corn	1612 (731)	—	1415 (642)	—
Ground soft wheat	—	1667 (756)	—	1460 (662)
Ground milo (grain sorghum)	—	—	—	—
Soybean meal solv. (50% protein)	265 (120)	190 (86)	302 (137)	236 (107)
Meat and bone meal (50% protein)	—	—	80 (36.3)	80 (36.3)
Alfalfa meal (17.5% protein)	40 (18.1)	40 (18.1)	40 (18.1)	40 (18.1)
Stabilized animal fat	—	30 (13.6)	—	20 (9.1)
Soybean oil	—	5 (2.3)	—	15 (6.8)
DL-methionine (98% methionine)	—	1.25 (0.57)	1 (0.45)	1.5 (0.68)
L-lysine (50% lysine)	—	1 (0.45)	—	—
Dicalcium phosphate (18.5% phosphate)	28 (12.7)	27 (12.3)	12 (5.4)	10 (4.54)
Limestone flour (38% calcium)	30 (13.6)	13.75 (6.25)	125 (56.7)	112.5 (51)
Iodized salt	5 (2.3)	5 (2.3)	5 (2.3)	5 (2.3)
Vitamin:mineral premix	20 (9.1)	20 (9.1)	20 (9.1)	20 (9.1)
Totals (lbs)	2000 (907)	2000	2000	2000
VITAMIN:MINERAL PREMIX				
Vit. A (millions of IU/ton)	5 (2.3)	6 (2.7)	8 (3.6)	9 (4.1)
Vit. D_3 (millions of ICU/ton)	0.8 (0.36)	0.8 (0.36)	1 (0.45)	1 (0.45)
Vit. E (thousands of IU/ton)	—	10 (4.5)	10 (4.5)	20 (9.1)
Riboflavin (g/ton)	3 (1.4)	3 (1.4)	6 (2.7)	6 (2.7)
Vit. B_{12} (mg/ton)	4 (1.8)	4 (1.8)	8 (3.6)	8 (3.6)
Niacin (g/ton)	30 (13.6)	30 (13.6)	50 (22.7)	50 (22.7)
d-calcium pantothenate (g/ton)	2 (0.9)	1 (0.45)	6 (2.7)	3 (1.4)

COMPLETE RATIONS FOR ADULT DUCKS
(PELLETED) *(CONTINUED)*

Ingredient	Ration 13 Corn Base Holding	Ration 14 Wheat Base Holding	Ration 15 Corn Base Breeder	Ration 16 Wheat Base Breeder
Choline chloride (g/ton)	100.0	—	300.0	—
Folic acid (g/ton)	—	—	0.5	0.5
Manganese sulfate (oz/ton)	9.6	9.6	9.6	9.6
Zinc oxide (80% zinc, oz/ton)	3.2	3.0	3.2	3.0
Ground grain to make 20 lbs	+	+	+	+
Totals (lbs)	20.0	20.0	20.0	20.0
CALCULATED ANALYSIS				
Crude protein (%)	14.2	13.6	16.3	15.8
Lysine (%)	0.63	0.60	0.77	0.72
Methionine (%)	0.26	0.26	0.33	0.33
Methionine + cystine (%)	0.50	0.53	0.58	0.58
Metabolizable energy (kcal/lb)	1419	1359	1322	1278
Calorie:protein ratio	100	100	81	81
Crude fat (%)	3.3	3.4	3.2	3.6
Crude fiber (%)	2.6	2.8	2.6	2.7
Calcium (%)	0.95	0.64	2.99	2.75
Available phosphorus (%)	0.35	0.36	0.40	0.39
Vit. A (IU/ton)	5240	4678	6610	6178
Vit. D$_3$ (ICU/ton)	400	400	500	500
Riboflavin (mg/lb)	2.17	2.21	3.73	3.76
Total niacin (mg/lb)	25	38	35	46

KEY: IU = international units
ICU = international chick units
g = gram
mg = milligram
Kcal = kilocalories

Symptoms of Vitamin and Mineral Deficiencies in Ducks

The symptoms for each deficiency below are listed in the order in which they usually manifest themselves.

VITAMINS

Biotin. Bottoms of feet are rough and callused, with bleeding cracks; lesions develop in corners of mouth, spreading to area around the bill; eyelids eventually swell and stick shut; slipped tendon (see also choline and manganese deficiencies); eggs hatch poorly

Sources: Most feedstuffs but especially dried yeast, whey, meat and bone meal, skim milk, alfalfa meal, soybean meal, green feeds. (The biotin in wheat and barley is mostly unavailable to poultry. If raw eggs are fed to animals, avidin — a protein in the egg white — binds biotin, making it unavailable.)

Choline. Retarded growth; slipped tendons (see also biotin and manganese deficiencies)

Sources: Most feedstuffs but especially fish meal, meat meal, soybean meal, cottonseed meal, wheat germ meal. (Evidence indicates that choline is synthesized by mature birds in quantities adequate for egg production.)

Folacin. Retarded growth; poor feathering; colored feathers show a band of faded color; reduced egg production; decline in hatchability; occasionally slipped tendon

Sources: Green feeds, fish meal, meat meal

Niacin. Retarded growth; leg weakness; bowed legs; enlarged hocks; diarrhea; poor feather development

Sources: Dried yeast and synthetic sources. (Most of the niacin in cereal grains is unavailable to poultry.)

Pantothenic acid. Retarded growth; viscous discharge causes eyelids to become granular and stick together; rough-looking feathers; scabs in corners of mouth and around vent; bottoms of feet rough and callused, but lesions are seldom as severe as in a biotin deficiency; drop in egg production; reduced hatchability of eggs; poor livability of newly hatched ducklings

    Sources: All major feedstuffs, particularly dried brewer's and torula yeasts, whey, skim and buttermilk, fish solubles, wheat bran, alfalfa meal

Riboflavin. Diarrhea; retarded growth; curled-toe paralysis; drooping wings; birds fall back on their hocks; eggs hatch poorly

    Sources: Dried yeast, skim milk, whey, alfalfa meal, green feeds

Thiamine. Loss of appetite; sluggishness; emaciation; head tremors; convulsions; head retracted over the back

    Sources: Grains and grain by-products

Vitamin A. Retarded growth; general weakness; staggering gait; ruffled plumage; low resistance to infections and internal parasites; eye infection; lowered production and fertility; increased mortality. (Adult birds develop symptoms more slowly than do ducklings.)

    Sources: Fish liver oils, yellow corn, alfalfa meal, fresh greens

Vitamin B_6. Poor appetites; extremely slow growth; nervousness; convulsions; jerky head movements; ducklings run about aimlessly, sometimes rolling over on their backs and rapidly paddling their feet; increased mortality

    Sources: Grains and seeds

Vitamin B_{12}. Poor hatchability; high mortality in newly hatched ducklings; retarded growth; poor feathering; degenerated gizzards; occasionally slipped tendon

    Sources: Animal products and synthetic sources

Vitamin D_3. Retarded growth; rickets; birds walk as little as possible, and when they do move, their gait is unsteady and stiff; bills become soft and rubbery and are easily bent; thin-shelled eggs; reduced egg production; bones of wings and legs are fragile and easily broken; hatchability is lowered

    Sources: Sunlight, fish liver oils, synthetic sources

Vitamin E. Unsteady gait; ducklings suddenly become prostrated, lying with legs stretched out behind, head retracted over the back; head weaves from side to side; reduced hatchability of eggs; high mortality in newly hatched ducklings; sterility in males and reproductive failure in hens

Sources: Many feedstuffs both of plant and animal origin, particularly alfalfa meal, rice polish and bran, distiller's dried corn solubles, wheat middlings

Vitamin K. Delayed clotting of blood; internal or external hemorrhaging, which may result in birds bleeding to death from even small wounds

Sources: Fish meal, meat meal, alfalfa meal, fresh greens

MINERALS

Note: While work on vitamin and mineral deficiencies in poultry has been limited largely to chickens and turkeys, I have observed many of these symptoms in ducks.

Calcium and phosphorus. Rickets; retarded growth; increased mortality; in rare cases thin-shelled eggs

Sources: Calcium — oyster shells and limestone. Phosphorus — dicalcium phosphate, soft rock phosphate, meat and bone meal, fish meal. (Most all feedstuffs have varying amounts of calcium and phosphorus, but typically only about one-third of the phosphorus in plant products is available to birds.)

Chloride. Extremely slow growth; high mortality; unnatural nervousness

Sources: Most feedstuffs, especially animal products, beet molasses, alfalfa meal.

Copper and iron. Anemia

Sources: Most feedstuffs, including fresh greens

Iodine. Goiter (enlargement of the thyroid gland); decreased hatchability of eggs

Sources: Iodine. (Iodized salt contains such small quantities of iodine that it cannot be relied upon to provide sufficient iodine.)

Magnesium. Ducklings go into brief convulsions and then lapse into a coma from which they usually recover if they are not swimming; rapid decline in egg production.

Sources: Most feedstuffs, especially limestone, meat and bone meals, grain brans. (Raising either the calcium or phosphorus content of feed magnifies a deficiency of this mineral.)

Manganese. Slipped tendon in one or both legs; retarded growth; weak eggshells; reduced egg production and hatchability. Slipped tendon (also known as perosis) is first evidenced by the swelling and flattening of the hock joint, followed by the Achilles tendon slipping from its condyles (groove), causing the lower leg to project out to the side of the body at a severe angle. (Perosis can also be the result of biotin or choline deficiencies.)

    Sources: Most feedstuffs, especially manganese sulfate, rice bran, limestone, oyster shell, wheat middlings, and bran

Potassium. Rare, but when it occurs, results in retarded growth and high mortality

    Sources: Most feedstuffs

Sodium. Poor growth; cannibalism; decreased egg production

    Sources: Most feedstuffs but especially salt and animal products

Zinc. Retarded growth; moderately to severely frayed feathers; enlarged hock joints; slipped tendon

    Sources: Zinc oxide and most feedstuffs, especially animal products

Summarized from Nutrient Requirements of Poultry, *7th revised edition (1977), pages 11–20, with the permission of the National Academy of Sciences, Washington, D.C.*

Predators

One of the most frustrating experiences for the owner of ducks is to lose valuable birds and eggs to predators. It is a good idea to talk with poultry keepers in your area to learn what kind of problems they have encountered and what precautions are necessary to minimize losses.

Raising a few domestic geese with your ducks can significantly reduce predation losses — especially during the day — in many situations. I frequently have said that I would not consider raising ducks here in this predator-rich location without the watchful presence of geese.

Of domestic animals, dogs — and to a lesser extent cats — are notorious for the damage they can inflict on poultry flocks. The first time one of these pets shows an interest in your ducks, they must be disciplined immediately if future troubles are to be avoided.

Some of the wild creatures that cause grief to duck raisers are rats, ground squirrels, weasels, mink, skunks, raccoons, opossum, foxes, coyotes, turtles, snakes, crows, ravens, magpies, jays, gulls, hawks, falcons, and owls. Because these animals are essential in maintaining the delicate balance of nature, we must not attempt to eliminate them but should respect them and give poultry sufficient protection so that hungry predators will not be able to dine at the expense of our birds.

Security Measures

Ducks roost on the ground, making them most vulnerable at night. Because many predators have nocturnal habits, a building or fenced yard where ducks can be comfortably locked in after dark is a must in most localities. If you raise ducks without penning them up nightly, it is almost 100 percent certain that your flock will be ravaged sooner or later.

A sturdy woven-wire fence with openings no larger than 2 × 4 inches at least 4 feet (1.25 m) tall is a starting point of keeping ducks safe while they sleep. In areas with determined hunters such as raccoons, skunks, bobcats, fox, mink, and opossum, electric fencing can be used in combination with woven-wire fences. To be the most effective, a minimum of three strands of electric wire should be installed on insulators mounted on the outside of the fence posts. One strand should be 3 to 4 inches (10 cm) above ground level, the second one approximately 9 inches (15 cm) above ground level, and the third strand several inches above the top of the woven wire fence. Additional strands can be used if extra security is desired. If weasels, mink, owls, and cats (wild or domestic) are prevalent in your area, pen your ducks at night in a shelter having a covered top as well as sturdy sides. Some predators will dig under fences or dirt floors of birdhouses. Burying the bottom 6 to 12 inches (15 to 30 cm) of fences that encircle yards or covering the floors of duck houses with wire netting will keep out excavators. PVC-coated wire lasts significantly longer than galvanized wire, especially if it is to be buried.

Because setting hens and ducklings are especially vulnerable to predation, extra care must be taken to provide them with secure quarters. Setting hens should be encouraged to nest in shelters that can be closed at night. When hens do nest in the open, panels 4 feet (1.25 m) high — or better yet, electroplastic fencing — can be set up around them to provide protection.

Identifying the Culprit

The following is a brief guide to help you recognize the work of some of the more common predators, with suggestions on how to stop their thievery.

Cats, domestic (feral or tame). *Telltale signs:* Crushed eggs held together by shell membranes. Birds disappear without a trace, or only a few feathers or clumps of down are found in a secluded spot where the animal fed. *Stopping losses:* Keep ducklings in wire-covered runs until they are four to eight weeks old, depending on the size of the breed. When a pet cat is seen stalking your birds, let the tabby know immediately that the ducks are off limits. Throwing a rolled-up newspaper at the offending animal will usually get the message across. If you have problems with stray cats, a live trap may be needed.

Cougars and bobcats. *Telltale signs:* Adult birds disappear with no trace or partially eaten carcass carefully covered and hidden with grass or other debris. *Stopping losses:* We have had good success stopping bobcats with a combination of either 4-foot or 5-foot-high woven wire and four strands of electrified smooth aluminum or galvanized steel wire mounted on insulators on the outside of this fence. The first electrified wire should be 2 to

3 inches (5 to 7.5 cm) above ground level, the second one approximately 9 inches (15 cm) above ground level, a third strand should be approximately 2½ feet (0.75 m) from the ground, and the fourth strand several inches above the top of the woven wire fence. **Caution:** For safety, always use an appropriate livestock fence–charger to energize electric fences! Cougars have been known to jump over 8-foot-tall deer fences, so the only sure protection is pens with covers strong enough to repel them.

Crows, gulls, jays, magpies, and ravens. *Telltale signs:* Punctured eggs or shells scattered around the base of an elevated perch such as a fencepost or tree stump. These birds also occasionally steal ducklings. *Stopping losses:* Provide covered nest boxes, and gather eggs several times a day. Keep ducklings in wire-covered runs until they are large enough to be safe — four to eight weeks depending on the size of the breed.

Dogs. *Telltale signs:* A number of ducks, sometimes the entire flock, badly maimed. Check for large holes under or through fences, and clumps of dog hair caught on wire. *Stopping losses:* Strong fences at least 4 feet (1.25 m) high.

Foxes. *Telltale signs:* Foxes are fastidious hunters and normally leave little evidence of their visits. Usually, they kill just one duck at a time and take the bird with them or partially bury it nearby. (However, foxes have been known to go on rampages, killing 30 or more birds at a time and scattering carcasses over a quarter-mile area.) You might be able to find a small hole under or through fences or a poorly fitted gate or door pushed ajar. *Stopping losses:* Tight fencing at least 4 feet (1.25 m) high with two strands of barbed or electric wire and close-fitting gates. Foxes will squeeze or dig under fences that are not flush with the ground.

Hawks and eagles. *Telltale signs:* Ducklings or small breed adult ducks disappear during daylight without a trace or only a few scattered feathers or clumps of down; larger breed ducks killed and partially eaten during daylight hours with the remains left where the bird was killed. Hawks generally will not take anything larger than 3 pounds, but eagles have been known to take medium-size geese. If you do occasionally lose fowl to hawks, remember that the average hawk eats 200 to 300 rodents yearly. *Stopping losses:* Keep ducklings in wire-covered runs until they are large enough to be safe. If large aggressive hawks such as Redtails are present, any duck that is approximately 3 pounds or smaller can be at risk if left in open-topped pens. Adult domestic geese often are a deterrent to hawks if the geese stay in close proximity to the ducks.

Mink and weasels. *Telltale signs:* Young ducklings disappear or larger birds are killed, evidently for amusement, with small teeth marks on head and neck. *Stopping losses:* At night lock the birds in a shelter that is covered with ½-inch (1.25 cm) wire hardware cloth. As incredible as it seems, mink and weasels can pass through holes as small as 1 inch in diameter.

Opossum. *Telltale signs:* Smashed eggs and birds that are badly mauled. *Stopping losses:* At night lock birds in a shelter with covered top and sides.

Owls, especially aggressive and bold, large species such as Great Horned. *Telltale signs:* One or more ducks killed nightly, usually with neck and chest (sometimes head) eaten. *Stopping losses:* At night lock ducklings and adult ducks in a completely covered shelter. Normally, pens covered with 1 × 1-inch (2.5 × 2.5 cm) wire netting will prevent losses if the enclosure/ pen is large enough that the owl cannot reach through and grab panicking birds. We have had Great Horned Owls kill ducks in temporary holding cages made with 1 × 2-inch welded wire. Some of the smaller species — such as screech owls – are mostly a threat to small ducklings and have been known to squeeze through 2 × 2-inch (5 × 5 cm) wire mesh.

Raccoons. *Telltale signs:* End of eggs bitten off, or crops (esophagus) eaten out of dead birds; possibly heads missing. Usually returns every fourth or fifth night. *Stopping losses:* Ideally, fowl are locked up from dusk until dawn every single night in a totally covered enclosure. At a minimum, the bottom 2 feet of the enclosure should be solid so the raccoon cannot reach through and grab frightened birds. The rest of the enclosure should have no openings larger than 2 × 4 inches. If a fenced yard without a top covering is used, the fencing must be tight along the bottom, at least 4 feet (1.25 m) high, have openings no larger than 2 × 4 inches, and have a minimum of one strand of electrified wire running 3 to 4 inches above ground level on the outside of the woven wire fence and a second strand several inches above the top of the fence. An uncovered fenced yard without the electrified strands will not deter raccoons. Raccoons are extremely persistent, strong, and intelligent, and will regularly return to see if any fowl are within reach.

Rats. *Telltale signs:* Eggs or dead ducklings pulled into underground tunnels. *Stopping losses:* Until ducklings are three to six weeks old, at night put them in a pen that has sides, top, and floor covered with ½-inch (1.25 cm) wire mesh. Rat populations should be kept under control with cats or traps.

Skunks. *Telltale signs:* Destroyed nest with crushed shells mixed with nest debris. *Stopping losses:* Gather eggs daily. Encourage setting hens to nest in shelters that can be at night, or set up panels — or better yet, electro-plastic fencing — around hens that are nesting out in open areas.

Snakes. *Telltale signs:* Nest appears untouched, but some or all eggs are missing. *Stopping losses:* Encourage setting hens to lay in covered nest boxes. A few geese can be effective in keeping snakes out of an area.

Snapping turtles and large fish. *Telltale signs:* Ducklings disappear mysteriously while swimming. *Stopping losses:* If your ducks frequent bodies of water that host turtles or large fish such as northern pike and largemouth bass, keep ducklings away from the water until they are two to four weeks old.

Duck Recipes

There are many ways duck can be fixed to taste great. Unfortunately, duck cookery is ignored or passed over lightly in most cookbooks. So with a lot of help from my wife, Millie, I'd like to share a few of our favorite recipes.

Steam-Fried Duck Eggs

Put eggs into medium-hot, lightly greased skillet. Pour a small amount of hot water around eggs and place lid on skillet. Steam-fry for several minutes until egg white is set. Serve immediately.

Roast Duckling

Serves 4 to 6

    1 whole 4½- to 5-pound. duckling
    1 tsp salt
    ⅛ tsp pepper
    3–4 cups homemade or 1 small box store-bought stuffing

Rub inside of duck cavity with salt and pepper. The duck may be roasted with or without stuffing. (In-bird stuffing often does not receive enough heat to cook thoroughly. To avoid the risk of illness from undercooking, make certain stuffing is cooked completely in bird or prepare stuffing as a side dish.) Fasten opening with skewers and lace closed with strong thread or thin string.

Place duckling uncovered on rack of roasting pan. Roast in preheated oven at 325°F (163°C) for 2½ to 3 hours, until skin is crisp and brown and flesh is tender.

Fried Duckling

Serves 4 to 6

1 duckling, cut into pieces

½ cup flour

1 tsp salt

⅛ tsp pepper

Dust pieces of duckling with flour. Season with salt and pepper. Fry in lightly greased skillet over medium-low heat for 1½ hours. Turn pieces as necessary to brown evenly. Remove fat from skillet as duck cooks, and use for gravy.

Duck Stew

Serves 6 to 8

2 cloves garlic, minced

2 Tbsp butter or margarine

1–2 cups leftover duck meat, in chunks

3 medium-sized potatoes, diced

4 carrots, sliced in ½-inch pieces

1 large onion, wedged

1 cup shredded cabbage

2 cups tomatoes

1–2 quarts vegetable or meat stock

Salt, pepper, paprika, and herbs, to taste

1 cup peas, green beans, or lima beans

2–3 celery stalks, cut in chunks

1 cup corn

In a large kettle sauté garlic in butter or margarine. Add the duck meat, potatoes, carrots, onion, cabbage, and tomatoes. Cover with vegetable or meat stock. Season with salt, pepper, paprika, and desired herbs.

Cook over medium-low heat until vegetables are crispy-tender. Add water if necessary to keep vegetables covered.

Add the peas or beans, celery, and corn. Heat to boiling point (or longer if fresh or frozen beans are used). If you prefer a thicker stew, add a paste of flour and water. Serve with fresh, warm homemade bread.

Here are two of our favorite recipes from our friends Gary and Kari Bennett:

Savory Roast Duck

Serves 4 to 6

3½- to 4½-pound duck (skin removed)

¾ tsp sage

¾ tsp rosemary

¾ tsp mace

¾ tsp paprika

¾ tsp garlic salt

¾ tsp onion salt

¾ cup mayonnaise

⅛ cup mustard

Combine the spices and the salts; rub onto duck.

Mix together mayonnaise and mustard (adjust proportions to taste). Generously coat seasoned duck with this mixture.

Bake in covered roaster at 300°F for a couple of hours (adjust for size of duck). Uncover for 30 minutes at end. Enjoy!

Curried Duck

Serves 4 to 6

2 to 3 strips bacon, diced

2½ to 3 pounds duck breast, cut into 1-inch pieces

½ to ¾ cup chopped onions

1 tsp curry powder

1 or 2 cans cream of mushroom soup

Potatoes or rice

Sauté a handful of bacon ends. Add duck breast pieces and onion.

Add curry to mixture and sauté until bonded. Mix in cream of mushroom soup and appropriate amount of water (according to directions on soup can).

Put in covered roaster and bake at 300°F for about 2 hours. Serve on potatoes or rice.

Whole-Wheat Angel Food Cake

Serves 10 to 12

1½ cups duck egg whites (8–10 eggs)

1½ tsp cream of tartar

¼ tsp salt

1 tsp vanilla

½ tsp almond flavoring

¾ cup packed brown sugar

1 cup sifted whole wheat flour

In a large bowl beat together (until stiff but glossy) the egg whites, cream of tartar, salt, vanilla, and almond flavoring.

Add the brown sugar to beaten whites, ¼ cup at a time, beating well after each addition. Fold in the flour with a large spoon, sifting a little over the top, folding in lightly with a down-up-over motion.

When well blended, pour into an ungreased 10-inch angel food cake pan. Bake at 375°F (190°C) for 45 to 60 minutes; touch the top gently to see if cake is done (it will spring back). Invert pan, and cool cake thoroughly. Remove from pan.

Option: Substitute ¼ cup carob or cocoa powder for an equal amount of flour, and omit almond flavoring.

Using Feathers and Down

Those soft feathers and down that were carefully saved at butchering time can be used to make a variety of handy items. While all ducks produce good-quality plumage, the down of Muscovies is less desirable than that of the breeds derived from Mallards. Large breeds such as Aylesbury, Pekin, and Rouen will produce 2½ to 3½ ounces (70 to 100 g) of down and small feathers per bird.

People frequently ask about plucking live ducks for down. While ducks usually survive such treatment, this practice places a great deal of stress on the birds and reduces their productivity. If ducks are live-plucked, taking the following precautions will reduce the negative effects it has on the birds: (1) Pick the birds only once a year during the late spring or early summer, after they have begun their natural molt. At this time, their feathers will be loosened, and the plumes can be removed with less pain to the birds. (2) Pull out only small pinches of feathers and down at a time to keep from tearing the bird's skin. (3) Take feathers only from the underside of the duck. (4) Remove a maximum of 50 percent of the feathers from the plucked area, and do not leave any bare patches. (5) Do not let the birds swim for 2 or 3 weeks, or until they have grown new feathers.

When making down pillows, a tightly woven cloth, such as downproof ticking, and double-stitched seams are essential to keep down and feathers from working their way out. Small, soft body feathers need to be mixed with down to make pillows more resilient. An excellent ratio is 75 percent down and 25 percent feathers, although good pillows can be made with a mixture of half down and half feathers.

A down-filled coverlet is not only delightful on a winter night but is also efficient. The tops and bottoms of comforters, quilts, or sleeping bags must be lined with downproof material. To keep the down and feathers distributed evenly, make channels 5 to 6 inches (13 to 15 cm) wide. Leave one end of the channels open for stuffing.

If the large plumes of the wings, tail, and body have been discarded at picking time, feathers and down can be used in the same ratio in which they come off the bird. As each channel is filled, sew the opening shut by hand, then finish the edges with binding.

For down clothing, as with comforters, you will need to make channels in order to keep the warmth evenly distributed. For parkas, vests, and other items, make the channels narrower, approximately 2 to 3 inches (5 to 7.5 cm) wide. Remember to use downproof lining and to double those seams so that none of the down escapes. To keep clothing as lightweight as possible, use 75 to 90 percent down, mixed with small quantities of body feathers.

APPENDIX F

Duck Breeders and Hatchery Guide

This guide is provided for your convenience and not as an endorsement of individual breeders.

Cackle Hatchery
Lebanon, Missouri
417-532-4581
www.cacklehatchery.com
Fawn & White Runner, Khaki Campbell, Mallard, Pekin, Rouen, Swedish

Eagle Nest Poultry
Oceola, Ohio
419-562-1993
www.eaglenestpoultry.com
Khaki Campbell, Mallard, Pekin, Rouen, Runner, Swedish

Hoffman Hatchery Inc.
Gratz, Pennsylvania
717-365-3694
www.hoffmanhatchery.com
Cayuga, Khaki Campbell, Mallard, Muscovy, Pekin, Rouen, Runner, Swedish, Welsh Harlequin

Holderread Waterfowl Farm & Preservation Center
Corvallis, Oregon
541-929-5338
www.holderreadfarm.com
Australian Spotted, Call, Cayuga, Dutch Hook Bill, East Indie, Khaki Campbell, Magpie, Miniature Silver Appleyard, Runner, Saxony, Silkie Duck, Silver Appleyard, Welsh Harlequin

Hoover's Hatchery
Rudd, Iowa
800-247-7014
www.hoovershatchery.com
Mallard, Pekin, Rouen

Ideal Poultry Breeding Farms, Inc.
Cameron, Texas
254-697-6677
www.idealpoultry.com
Cayuga, Crested, Fawn & White Runner, Khaki Campbell, Mallard, Orpington (Buff), Pekin, Rouen, Swedish

Marti's Hatchery
Windsor, Missouri
660-647-3156
www.martipoultry.com
Khaki Campbell, Mallard, Pekin, Rouen, Runner

Metzer Farms
Gonzales, California
800-424-7755
www.metzerfarms.com
Cayuga, Crested, Hybrids, Khaki Campbell, Mallard, Orpington (Buff), Pekin (various strains), Rouen, Runner, Swedish

Mt. Healthy Hatcheries
Mt. Healthy, Ohio
800-451-5603
www.mthealthy.com
Mallard, Pekin

Murray McMurray Hatchery
Webster City, Iowa
800-456-3280
www.mcmurrayhatchery.com
Cayuga, Crested, Runner, Khaki Campbell, Mallard, Orpington (Buff), Pekin, Rouen, Swedish

Privett's Hatchery, Inc.
Portales, New Mexico
800-634-4390
www.privetthatchery.com
Cayuga, Crested, Khaki Campbell, Mallard, Orpington (Buff), Pekin, Rouen, Runner, Swedish

Ridgeway Hatcheries
LaRue, Ohio
800-323-3825
www.ridgwayhatchery.com
Mallard, Muscovy, Pekin, Rouen, Runner, Swedish

Sand Hill Preservation Center
Calamus, Iowa
563-246-2299
www.sandhillpreservation.com
Ancona, Campbell (Dark, Khaki), Cayuga (Black, Blue), Crested, Golden Cascade, Magpie, Orpington (Buff), Pekin, Runner, Saxony, Silver Appleyard, Swedish, Welsh Harlequin

Strombergs' Chicks & Gamebirds Unlimited
Pine River, Minnesota
800-720-1134
www.strombergschickens.com
Aylesbury, Call (Gray, Snowy, White), Cayuga, Crested, East Indie, Runner, Khaki Campbell, Mallard, Muscovy, Rouen, Silver Appleyard, Swedish (Black, Blue), Welsh Harlequin

Sun Ray Chicks Hatchery
Hazelton, Iowa
319-636-2244
www.sunrayhatchery.com
Pekin, Rouen

Townline Hatchery
Zeeland, Michigan
888-685-0040
Pekin, Rouen

Welp, Inc.
Bancroft, Iowa
800-458-4473
www.welphatchery.com
Cayuga, Crested, Runner, Khaki Campbell, Mallard, Orpington (Buff), Pekin, Rouen, Swedish

APPENDIX G

Sources of Supplies and Equipment

Sources in this appendix are given for your convenience, not as an endorsement.

BF Products Inc.
Harrisburg, Pennsylvania
800-255-8397
www.bfproducts.com
Fencings and nettings

Brinsea Products Inc.
www.brinsea.co.uk
Small incubators

Brower Manufacturing Co.,
Hawk Eye Steel Products, Inc.
Houghton, Iowa
319-469-4141
www.hawkeyesteel.com
Brooders, waterers, feeders

Collapsible Wire Products, LLC
North Prairie, Wisconsin
262-781-6125
www.collapsiblewireproducts.com
Wire exhibition cages

Cutler Supply
Applegate, Michigan
810-633-9450
www.cutlersupply.com
Bands, medications, books, incubators, and other supplies

First State Veterinary Supply
Parsonsburg, Maryland
800-950-8387
www.firststatevetsupply.com
Medications

Gey Band and Tag Co., Inc.
Norristown, Pennsylvania
610-277-3280
Bands, tags, and other supplies

GQF Manufacturing Company
Savannah, Georgia
912-236-0651
www.gqfmfg.com
Incubators, brooders, supplies

J. A. Cissel Mfg. Co.
Lakewood, New Jersey
800-631-2234
http://www.jacissel.net
Netting, grid flooring, rigid fencing, catch nets, and other supplies

Jeffers, Inc.
Dothan, Alabama
800-533-3377
www.jefferspet.com
Medications and general supplies

Kalglo Electronics Co., Inc.
Bethlehem, Pennsylvania
888-452-5456
www.kalglo.com
Infrared brooder heaters

Kuhl Corporation
Flemington, New Jersey
908-782-5696
www.kuhlcorp.com
Incubators, feeders, waterers

Max-Flex
Lindside, West Virginia
800-356-5458
www.maxflex.com
Fences and netting

NASCO
Fort Atkinson, Wisconsin
800-558-9595
www.eNASCO.com
General farm supplies

National Band and Tag Co.
Newport, Kentucky
859-261-2035
www.nationalband.com
Bands and tags

NatureForm Hatchery Systems
Jacksonville, Florida
904-358-0355
www.natureform.com
Incubators

PMI Nutrition International
St. Louis, Missouri
800-227-8941
www.mazuri.com
Mazuri waterfowl feeds for all ages

Premier1 Supplies
Washington, Iowa
800-282-6631
www.premier1supplies.com
Fencing and netting

Smith Poultry & Game Bird Supply
Bucyrus, Kansas
913-879-2587
www.poultrysupplies.com
Medications, books, incubators, and
other supplies

**Stromberg's Chicks and
Gamebirds Unlimited**
Pine River, Minnesota
800-720-1134
www.strombergschickens.com
Bands, medications, books, incubators,
Dux-Wax, and other supplies

APPENDIX H

Organizations and Publications

General Organizations

American Bantam Association
Augusta, New Jersey
www.bantamclub.com
Publisher of the *Bantam Standard*

American Poultry Association
Burgettstown, Pennsylvania
secretaryapa@yahoo.com
www.amerpoultryassn.com
Publisher of the *Standard of Perfection*

Heritage Breed Organizations

American Livestock Breeds Conservancy
Pittsboro, North Carolina
919-542-5704
www.albc-usa.org

Rare Breeds Canada
Castleton, Ontario
905-344-1026
www.rarebreedscanada.ca

Rare Breeds Survival Trust
Warwickshire, United Kingdom
+44-0-24-7669-6551
www.rbst.org.uk

Society for the Preservation of Poultry Antiquities
Calamus, Iowa
319-246-2299
www.feathersite.com/Poultry/SPPA/SPPA.html

Publications

Fancy Fowl
+44-0-17-2862-2030
www.fancyfowl.com
Monthly color magazine for poultry and waterfowl hobbyists

Feather Fancier
519-542-6859
www.featherfancier.on.ca
Monthly newspaper and source for clubs, shows, and suppliers

Game Bird and Conservationists' Gazette
memberservices@gamebird.com
www.gamebird.com
Bimonthly magazine and source for suppliers of wild ducks and equipment

Poultry Press
765-827-0932
www.poultrypress.com
Monthly newspaper and source for clubs, shows, and suppliers

Glossary

air cell. A pocket of air that develops between the two shell membranes in the large end of eggs shortly after they are laid.

bean. The hard nail at the tip of a duck's bill.

breed. A subdivision of the duck family whose members possess similar body shape and size and the ability to pass these characteristics on to their offspring.

breed true. When offspring resemble their parents.

breeder ration. A feed used for the production of hatching eggs.

broiler. Quick-growing poultry used for meat at an early age.

buffled. Having the fluffy head feathering desired in Call ducks.

candling. The illumination of an egg's contents with a bright light.

caruncles. The bumpy flesh on the face of Muscovies.

class. The duck breeds of similar size that are grouped together in the *American Standard of Perfection*.

cloaca. The inner cavity behind the vent where the urinary, intestinal, and reproductive canals open in fowl.

concentrated feed. Feeds that are high in protein, carbohydrates, fats, vitamins, and minerals and low in fiber.

crest. The elongated feathers on the head of some breeds of ducks.

down. The furlike covering of newly hatched ducklings.

drake. A male duck.

drakelet. A young male duck.

duck. In general any member of the Anatidae family; it is often used specifically in reference to females of the duck family.

ducklet. A young female duck.

ducklings. Young ducks up to the point where their feathers have completely replaced their baby down.

eclipse molt. A 3- to 4-month period each year, after the breeding season, when the bright plumage of colored adult drakes is replaced with subdued colors similar to those of females.

egg tooth. Small, horny protuberance attached to the bean of the bills of newly hatching birds that is used to help break the shell at hatching time. It falls off several days after hatching.

fault. Any characteristic of a purebred duck that falls short of the written standard.

feed conversion. The ability of birds to convert feed into body growth or eggs. To calculate feed conversion ratios, divide pounds of feed consumed by pounds of body weight or eggs.

fertile. Sexually mature ducks that are capable of reproducing; also, an egg that contains a viable embryo.

fertility. In reference to eggs, the capability of producing an embryo. Fertility is expressed as a percentage that equals the total number of eggs set minus those that are infertile divided by the total number set times 100.

flights. The large feathers of the wing, including the primaries and the secondaries.

flock mating. A group of breeding ducks that are penned together and consist of more than one male and two females.

green ducklings. Young ducks that are managed for fast growth and then slaughtered at seven to eight weeks of age.

growing ration. A feed that is formulated to stimulate fast growth in ducklings over two weeks old.

hatchability. The ability of eggs to hatch. Hatchability can be expressed as (1) a percentage of the fertile eggs set (total number of ducklings hatched divided by the number of fertile eggs set times 100) or (2) a percentage of all eggs set (total number of ducklings hatched divided by the total number of eggs set, times 100).

hybrid. The offspring of a planned breed cross; also, the offspring of parents of different species.

incubation period. The number of days it takes eggs to hatch once they are warmed to incubation temperature.

infertile. Eggs that do not contain an embryo; also, mature animals that are incapable of reproduction.

keel. The pendulous fold of skin hanging from the underbody of some ducks.

laying ration. Feed that is formulated to stimulate high egg production.

linebred. A planned form of inbreeding that conserves the genetic material of specific ancestors.

maintenance ration. A feed used for adult ducks that are not in production.

molt. The natural replacement of old feathers with new ones.

nuptial plumage. In colored varieties of ducks, the bright breeding plumage of males exhibited during fall, winter, and spring.

oil gland. Also known as uropygial gland. Located at the base of the tail, it produces water-repellent oil for the feathers.

old drake or old duck. A drake or duck more than one year old.

oviduct. The tube that transports the egg from the ovary to the cloaca.

pair. A breeding unit consisting of a male and a female.

pinfeathers. New feathers that are just emerging from the skin.

pip. The first visible break the duckling makes in the eggshell.

primaries. The ten large, outermost feathers of each wing.

production-bred. Ducks that have been selected for top meat or egg production.

pullet. A young fowl in her first year of egg production.

purebred. A duck having parents of the same breed.

ration. The feed consumed during the course of a day.

roached back. A deformed, twisted, or humped back.

scoop-bill. A bill with a topline that is severely concaved.

secondaries. The flight feathers growing on the inner half of the wing.

sex feathers. The curled feathers in the center of the tail when Mallard or Mallard-derived drakes are in nuptial plumage.

sport. A colloquial term for mutation.

standard-bred. Ducks that have been stringently selected over many generations according to the ideal that is set forth in the *Standard of Perfection*.

Standard of Perfection. A book containing pictures and descriptions of the physical characteristics desired in the perfect bird of each recognized breed and variety of poultry.

starting ration. A high-protein feed used during the first couple of weeks to get ducklings off to a good start.

straight run. Young poultry that have not been sexed.

strain. A group of animals within a breed that are more closely related than the general population of that breed.

strain-cross. The mating together of males from one strain to females of a second strain.

tertials. The band of large wing feathers growing next to the body.

trio. A breeding unit consisting of a male and two females.

type. The unique shape and size shared by animals of the same breed.

variety. A subdivision of the breeds. In ducks the varieties within a breed are identified by their plumage color or markings.

waterfowl. Birds that naturally spend most of their lives on and near water. This term is often used in specific reference to ducks, geese, and swans.

wingbow. The top surface of the wing, encompassing the area between the wings leading edge and the speculum coverts, and between the wing shoulder to the primary coverts.

young drake or duck. A drake or duck less than one year old.

Index

Page references in *italics* indicate illustrations;
references in **bold** indicate charts or graphs.

A

abdomen, distended, 279–80
acquiring stock, 157–61
  day-old ducklings, 158–59
  hatching eggs, 158
  how many to buy, 160–61
  mature, 160
  production-bred vs. standard-bred
    stock, 157
adult duck management, 215–48
  basics, 215–16
  breeders/breeding, 228, 243–44
  catching/holding ducks, *245,*
    245–47
  clipping wings, 247, *247*
  drinking water, 226–27, 229
  housing, 216–26
  laying flock, 237–38
  manure management, 248
  nutrition for, 229–38
  salt water, raising ducks on, 229
  sexing of mature ducks, 247–48
  swimming water, 226
American Bantam Association,
  26–27
American Poultry Association, 26
American Standard of Perfection, 27

anatomy, 13–16, *14*
  back, 16
  bill, 15, *15*
  body shape, 13
  deformities/weaknesses, 243–44
  eyes, 15
  feathers, 14
  feet and legs, 16
  keel, 16
  tail, 15
  wings, 14
Ancona, 68–69, *69,* 103
antibiotics, 255, 271
Appleyard, 80–82, *81*
  breeders, selecting, 81–82
  comments, 82
  description of, 80–81
  show birds, 81–82
  varieties, 81
Appleyard, Reginald, 46, 80
artificial incubation, 167–71
  leveling the incubator, 169
  multistage/single-stage, 169
  operating specifications, 169
  placement of incubator, 168
  step-by-step, 170–71
  types of incubators, 168

STOREY'S GUIDE TO RAISING SERIES

For decades, animal lovers around the world have been turning to Storey's classic guides for the best instruction on everything from hatching chickens, tending sheep, and caring for horses to starting and maintaining a full-fledged livestock business. Now we're pleased to offer revised editions of the Storey's Guide to Raising series — plus one much-requested new book.

Whether you have been raising animals for a few months or a few decades, each book in the series offers clear, in-depth information on new breeds, latest production methods, and updated health care advice. Each book has been completely updated for the twenty-first century and contains all the information you will need to raise healthy, content, productive animals.

Storey's Guide to Raising BEEF CATTLE (3rd edition)

Storey's Guide to Raising RABBITS (4th edition)

Storey's Guide to Raising SHEEP (4th edition)

Storey's Guide to Raising HORSES (2nd edition)

Storey's Guide to Training HORSES (2nd edition)

Storey's Guide to Raising PIGS (3rd edition)

Storey's Guide to Raising CHICKENS (3rd edition)

Storey's Guide to Raising DAIRY GOATS (4th edition)

Storey's Guide to Raising MEAT GOATS (2nd edition)

Storey's Guide to Raising DUCKS (2nd edition)

Storey's Guide to Raising MINIATURE LIVESTOCK (NEW!)

Storey's Guide to Keeping HONEY BEES (NEW!)

Storey's Guide to Raising TURKEYS

Storey's Guide to Raising POULTRY

Storey's Guide to Raising LLAMAS

Other Storey Titles You Will Enjoy

The Backyard Homestead, edited by Carleen Madigan.
A complete guide to growing and raising the most local food available anywhere —
from one's own backyard.
368 pages. Paper. ISBN 978-1-60342-138-6.

How to Build Animal Housing, by Carol Ekarius.
An all-inclusive guide to building shelters that meet animals' individual needs:
barns, windbreaks, and shade structures, plus watering systems, feeders, chutes,
stanchions, and more.
272 pages. Paper. ISBN 978-1-58017-527-2.

Keep Chickens!, by Barbara Kilarski.
Everything you need to know to raise healthy chickens in small urban or suburban
environments.
160 pages. Paper. ISBN 978-1-58017-491-6.

Keeping a Nature Journal, by Clare Walker Leslie & Charles E. Roth.
Simple methods for capturing the living beauty of each season.
224 pages. Paper with flaps. ISBN 978-1-58017-493-0.

Made from Scratch, by Jenna Woginrich.
A memoir written by a part-time homesteader, sharing the honest satisfaction of
learning forgotten skills and always finding the humor and joy in the journey.
208 pages. Paper. ISBN 978-1-60342-532-2.
Hardcover with jacket. ISBN 978-1-60342-086-0.

Outdoor Water Features, by Alan and Gill Bridgewater.
Attractive plans for introducing pools, ponds, cascades, and more to your yard and
garden.
128 pages. Paper. ISBN 978-1-58017-334-6.

Pocketful of Poultry, by Carol Ekarius.
More than 100 amazing poultry pals in full-page color photographs.
272 pages. Paper. ISBN 978-1-58017-677-4.

Storey's Illustrated Guide to Poultry Breeds, by Carol Ekarius.
A definitive presentation of more than 120 barnyard fowl, complete with full-color
photographs and detailed descriptions.
288 pages. Paper. ISBN 978-1-58017-667-5.
Hardcover with jacket. ISBN 978-1-58017-668-2.

These and other books from Storey Publishing are available
wherever quality books are sold or by calling 1-800-441-5700.
Visit us at *www.storey.com.*